The gathering and analysis of experimental data are fundamental activities in science and engineering. *Mathematica® in the Laboratory* is a hands-on guide which shows how to harness the power and flexibility of *Mathematica* in the control of data-acquisition equipment and the analysis of experimental data. It is fully compatible with *Mathematica* 3.0.

The book is made up of two parts. The first explains how to use *Mathematica* to import, manipulate, visualize and analyze data from existing files. The generation and export of test data are also covered. The second part deals with the control of laboratory equipment. The use of *Mathematica's* *MathLink®* system as applied to instrument control, data processing, and interfacing is clearly demonstrated.

Many practical examples are given, which can either be used directly or adapted to suit a particular application. The book sets out clearly how *Mathematica* can provide a truly unified data-handling environment, and will be invaluable to anyone who collects or analyzes experimental data, including astronomers, biologists, chemists, mathematicians, geologists, physicists, and engineers.

Mathematica® in the Laboratory

Mathematica® in the Laboratory

Samuel Dick
Royal Greenwich Observatory

Alfred Riddle
Macallan Consulting, California

Douglas Stein
Addison Wesley Interactive

CAMBRIDGE
UNIVERSITY PRESS

PUBLISHED BY THE PRESS SYNDICATE OF THE UNIVERSITY OF CAMBRIDGE
The Pitt Building, Trumpington Street, Cambridge CB2 1RP, United Kingdom

CAMBRIDGE UNIVERSITY PRESS
The Edinburg Building, Cambridge CB2 2RU, United Kingdom
40 West 20th Street, New York, NY 10011-4211, USA
10 Stamford Road, Oakleigh, Melbourne 3166, Australia

First published 1997

Printed in the United States of America

Typeset in Palatino

Library of Congress Cataloging-in-Publication Data

Dick, Samuel.
Mathematica® in the laboratory / Samuel Dick, Alfred Riddle,
Douglas Stein.
p.   cm.
Includes bibliographical references.
ISBN 0 521 49906 2
1. Automatic data collection systems – Computer programs.
2. Mathematica (Computer file)   I. Riddle, Alfred.   II. Stein,
Douglas.   III. Title.
TS158.6.D53   1997
502'.85'53 – dc20   96-21448
                                                    CIP

*A catalog record for this book is available from
the British Library*

ISBN   0 521 58137 0 hardback
ISBN   0 521 49906 2 paperback

To Alison and my parents
Samuel Dick

To Dad, whose workshop was my first lab, to P. D. Evans, who taught me that a good experimentalist is a physicist at heart, and to R. J. Trew, who said, "If you'll run my lab, I'll make sure no one touches your setups."
Alfred Riddle

To Susan, Amanda, and Joel, who had to leave me at home on weekends to work on this book. Also to my Mom and Dad, who allowed me to fire up the soldering iron and build radios and other gizmos in my pre-teen years.
Douglas Stein

Contents

Preface

Mathematica is a powerful tool to have in your laboratory. But what do we mean by laboratory, and what exactly is a powerful tool? The type of laboratory that is uppermost in our minds will have ongoing experiments, instruments with displays and dials, and a computer. Perhaps the laboratory has rows of bottles with colored liquids and Petri dishes. Or it might instead have growing plants, an environmental chamber for testing a new solidstate device from a silicon foundry, or a telescope collecting light from a distant galaxy. The laboratory itself might even be an entire state, a country, or an ocean, and the data might be gathered by surveying aircraft, by satellite, or by radio telemetry from remote helicopter-supported ground sites.

Whenever there is both data gathering and data processing to be done, we believe you will find *Mathematica* powerful. Why? Because *Mathematica* can provide you with a "unified" environment within which data visualizing, symbolic mathematical modeling, instrument control, data acquisition, data analysis, and report generation are all possible.

Traditionally, engineers and scientists have required different software packages for instrument control, for data acquisition, and for analysis. For example, an instrument control program would setup and acquire data from, say, a digital voltmeter which is connected by some means to a computer. The control program would place data into a file on the computer's disk. Another program would open that file, perhaps to graphically display the data, to compare the data with computer-generated data from a theoretical model, or to execute some statistical test. In our experience, it is not uncommon for several programs (some written by us) to be used to process only one set of data – because each program is limited in its functionality. We have all spent a lot of time transferring numbers between programs, not to mention learning how to make many pieces of software do exactly what we need.

In this book we show you how you can do all these tasks – from instrument set-up to complicated analysis – by using *Mathematica*. (For instrument control, you will also need to be able to create short programs in the high-level language C. The focus of your programming work, however, will always be at a high level: the assembler manual can stay firmly on the shelf.)

We believe *Mathematica* forms an ideal platform for laboratory work. It is flexible and comprehensive enough to allow us to analyze, plot, and present our data. Its ASCII file format is an added blessing because we can always extract our data from the *Mathematica* file. The power and flexibility of the *Mathematica* language may seem daunting at first, but very quickly we find that this power keeps us from running into frustrating – and time-consuming – limitations. The only problem to date for a laboratory user is that *Mathematica* lacks data acquisition functions. In response, this book provides data acquisition solutions for *Mathematica* and gives many examples and routines for data analysis and plotting with *Mathematica*. We hope *Mathematica* saves you time, increases your insight, and helps you explain your findings as much as it does for us.

With *Mathematica*, your choice of computer is also wide open. *Mathematica* is available on a wide range of computers: IBM PCs (running DOS, Windows, OS/2, or Linux) and RISC workstations, the Apple Macintosh series (including Power Macintosh), Sun systems, DEC Alpha OSF and RISC Ultrix workstations, Hewlett-Packard workstations, and Silicon Graphics machines – to name just the most common computers. Across all of these computers, *Mathematica* has a very consistent user interface. This consistency minimizes your learning time when you move between machines, or that of a colleague when you pass your work on.

But why write a book about how to use *Mathematica* in the laboratory? *Mathematica* is a large and many-featured software system. Our aim with this book is to show you time-saving direct routes to solutions in typical scientific and engineering application areas. Of course, we encourage you to explore as much of *Mathematica* as you can. Our book can only guide you so far.

It is not possible within the scope of any one book to cover all the possibilities and all the problems that everyone will encounter. We try to provide an overview of work methods and to show in a general way how we develop those methods into solutions. That is, we concentrate on describing and discussing the foundation techniques that you need to know in order to develop your own solutions, targeted to your needs. By expanding upon our examples, you will use one of *Mathematica*'s great strengths: its customizable extensibility.

Structure

We have written the book in two main sections. In both sections it is our aim to provide a foundation layer of techniques that you will be able first to apply immediately and then to augment, once your familiarity with *Mathematica* grows.

In the first five chapters, we show how you can use *Mathematica* to import data from existing files, to graph data, to fit functions to data, and to generate and export test data. A fresh start is not necessary: you can gain from using *Mathematica* with your existing archival data. In the later chapters, we discuss both the task of controlling laboratory equipment and *Mathematica*'s *MathLink* system which enables you to extend *Mathematica* into the area of instrument control or to write your own data processing functions. We show how you can control instruments using a serial link and a general interface card – all driven from *Mathematica*.

Please note that it is not necessary to read the book chapter by chapter. Dip in, browse through, and skip over as you like. The material in any one chapter is not necessarily related to the material in any other chapter.

Conventions

Because we use *Mathematica* interactively, we have preceded our instructions to *Mathematica* by *In:* (and typeset them in **bold monospace font**). *Mathematica*'s subsequent reply is preceded by *Out:* (in `plain monospace font`). Where *Mathematica* generates graphical output, it follows immediately after the generating input; where *Mathematica* generates sound, it is marked -Sound-. We have set examples of non-*Mathematica* code and references to file names in `plain monospace font`, too.

Mathematica 2.2 and 3.0

When this book is published, 3.0 will be the newest version of *Mathematica*. We appreciate that many folks will still be using version 2.2. In general, *Mathematica* 2.2 and 3.0 both work happily with our examples. Where we have found significant differences between the *Mathematica* versions, we have added the version number as a suffix to the *In:* and *Out:* indicators described above. For example, *In 3.0:*. Where we found no significant difference, we have used version 2.2 input and output formats. We have verified our code using the last *beta* version of 3.0 – it is, however, possible that some minor differences may exist between this *beta* version and the first commercially-released version.

Electronic access to material

We have placed the example datafiles and extracts of the *Mathematica* and C code on *MathSource*, Wolfram Research's electronic resource for *Mathemat-*

ica material. You can access *MathSource* by mail from the Internet, or from the Wolfram Research website.

Acknowledgments

The authors thank the many people who have helped us write this book. Stephen and Conrad Wolfram, Prem Chawla, and Todd Gayley at Wolfram Research are thanked for useful conversations and for answering our puzzles.

The team at Cambridge University Press, especially David Tranah in Cambridge and Philip Meyler in New York, are thanked for their enthusiasm, their encouragement and support, and their patience.

DBS wishes to acknowledge Shawn Sheridan of Wolfram Research for many helpful discussions over the years on the design of *MathLink* and the proper design of *MathLink*-based applications. Without his painstaking work, the C-language interfaces for *MathLink* would not be the elegant and useful tools that they are. DBS also wishes to thank and acknowledge Rob Dye and Ed McConnell of National Instruments, who graciously arranged for the loan of the LAB-LC board and copy of the NI-DAQ libraries used in the examples of Chapter 9.

Samuel Dick
Alfred Riddle
Douglas Stein
1996

CHAPTER 1

Importing data from files

Most experimentally biased scientists and engineers carefully store the data resulting from past work. These archived data may be stored in many different formats. Indeed, data that you are collecting now may be in several different formats because it comes from different experiments that use different data acquisition techniques, software, and computer hardware. If you analyze these data with a system other than that with which they were acquired, chances are that you will need to perform some sort of conversion. You will certainly need to "import" those data into whichever system you are using.

Your *Mathematica* system is no exception to this general rule, but the task of reading in, or importing, data from non-*Mathematica* files is really rather easy. Once the data are in *Mathematica*, you will be able to use all of *Mathematica*'s functionality to graph and to analyze them – perhaps exporting the data at a later date. *Mathematica* gives you considerable flexibility in how you handle data. For example, if your data have a particular structure, you will be able to retain that structure, even if it contains a mix of numeric data types and text. If you are used to dealing with mixed-type structures in C (or records in Pascal) then you will be able to build upon your experience with such data groupings when you work with *Mathematica*.

In this chapter we show you how to import data from ASCII and binary files; we also show you how, in general terms, to extract data that may be embedded amidst other, nonpertinent, contents of a file, and to maintain any structure that might be present. First, we begin by showing you how to make *Mathematica* navigate around the filesystem of your computer.

1.1   File operations in general

Before you can import data from a file, you need to tell *Mathematica* where that file is to be found. All computer systems operate some kind of filing system. At the top-most level, the filing system will use as the first element of a file's name the name of the disk on which the data are stored. For example, the computer on which this is being written has three disks attached, named `Macintosh HD`, `Q127 Alpha`, and `Q127 Beta`; the latter two disks are actually one physical hard disk that is partitioned into two logical disks by the operating system – so the computer sees two independent disks. Such physical partitioning does not affect what we are about to show: it is the name (or symbol) of the disk on which your data are resident that is significant.

Within each disk there will be some kind of lower-level partitioning. A disk on a multi-user system will have directories, one for each user. A disk on a single-user machine (or a user's directory on a multi-user machine) also will have directories or folders. These, too, may contain many sets of nested subdirectories. Typically, each level of directory structure is delineated by some special character like \, /, or :, depending on which operating system is in use – DOS, UNIX, or Macintosh respectively.

Before you begin to read and write files using any applications software, you should feel confident about how your computer system defines the disk and directories that contain a file, and also what filenames are allowed. For example, you should know how many characters you can use in a filename, whether you are disallowed certain characters (like spaces or colons) within a filename, and which character is used to indicate a new level within the filing system's directory structure. We suggest also that you try out new techniques only on files for which you have backup copies – which preferably are stored away on media not connected to your computer.

1.1.1 Locating files

At any time during a *Mathematica* session, *Mathematica* has a default directory for ongoing work. By default, it is this directory that *Mathematica* will access for all file operations. You can see the name of the default directory by using the **Directory** function, here shown for a Macintosh system in which filenames can have spaces, and in which levels within the file system are delineated by colons:

In:

```
Directory[]
```

Out:

```
Q127Beta:Mathematica 2.2
```

As returned by **Directory**, the current default directory is called `Mathematica 2.2` and is located on a disk called `Q127Beta`. If the default directory happens to contain the file you wish *Mathematica* to read, then you need go no further. If the file is elsewhere, then you have three options. Either you can set the default directory to be the directory that contains the file, you can explicitly specify the full filename whenever you want to access the file, or you can add the name of the directory in which the file is located to *Mathematica*'s **$Path** variable. **$Path** is a list of directories through which *Mathematica* will search in order to find a file. For example, if you know the directory name, then you can use the **SetDirectory** function to make *Mathematica* use that directory as the default. The **FileNames** function will then list all the files within that directory.

In:

```
SetDirectory["Q127Alpha:Sam's observations"]
```

Out:

```
Q127Alpha:Sam's observations
```

In:

```
FileNames[]
```

Out:

```
{First experiment, noGood.dat}
```

Note that the filenames within the chosen directory have been returned as a list, enclosed in curly brackets, like all *Mathematica* lists. If you get the directory name wrong, *Mathematica* will warn you with a message like

Out:

```
SetDirectory::cdir: Cannot set current directory to
Q127Alpha:Sam's experiment.
```

If you are going to work often on a file or on a group of files that will always be resident in one particular directory, adding that directory to *Mathematica*'s default search path can save you time. You can see which device and directory combinations are already on the default search path by just typing **$Path**.

In:

```
$Path
```

Out:

```
{Q127Beta:Mathematica 2.2:Packages,
Q127Beta:Mathematica 2.2:Packages:StartUp, :}
```

You can append your chosen directory's name to the search path using the **AppendTo** function.

In:

```
AppendTo[$Path,"Q127Alpha:Sam's observations"]
```

Out:

```
{Macintosh HD:Mathematica 2.2.2:Packages,
Macintosh HD:Mathematica 2.2.2:Packages:StartUp, :,
Q127Alpha:Sam's observations}
```

Now, *Mathematica* will automatically search the directory Sam's observa-tions on the disk Q127Alpha. (You can also use the **PrependTo** function to make the new directory name the first element in **$Path**.) Once you have established that *Mathematica* can access the directory containing the required file, you can proceed to make *Mathematica* read it.

In addition to knowing where a file is located in your computer's filesys-tem, you also need to know the type of file and how the data are structured within the file.

1.2 File types

Data are stored typically in one of two main types of file: ASCII or binary. In ASCII files, data are stored in a printable format using the ASCII code; data in binary files are store in base-2 form. The advantage of using ASCII data files is that usually you can type or print the files to see what is in them. Accessing data in binary files is not so simple because you need to know how the data were stored in the files – owing to the different ways that computers store integers and floating point numbers, just to take two examples. Trying to type or print a binary file can produce strange results – including "random" flashing of the terminal screen and multiple formfeeds from printers – because the device to which the contents of the file are being sent will interpret some of the binary values as device control codes, which are interpreted as instructions for the device to behave in a particular way. However, binary files normally occupy less space, and so they are especially useful for large sets of data. *Mathematica* can read and write both binary and ASCII files with ease.

If your data are contained in multiple files, we show you later in the chapter how you can use *Mathematica* to access all the files automatically, or even how to use the data themselves to specify which files have to be accessed next or how the data are to be stored.

1.3 Data structures

Before reading in any data, you will find it useful to know both how the data are arranged in the file to be read and how you want to use the data after it is read into *Mathematica*. The former is essential; the latter is merely desirable. If the data are in some kind of structure, or are related in some way, then you might want to keep that structure or relationship. In C or Pascal, such structures are given special names like `struct` and `record`. *Mathematica* handles data and structures in a more general way: all groups of data are lists, and a list can contain identical or differently typed members.

For example, a chronological date might consist of three numbers – the year, the month, and the day. In *Mathematica*, you can keep these three quantities within a list. Here is a list (containing a date) that has been assigned to a variable name **myDate**:

In:
```
myDate={1900,1,1}
```
Out:
```
{1900, 1, 1}
```

Of course, you might want to mix data of different types within this date structure. For example, a different format of date might contain two integers and a string.

In:
```
myDate2={1900,"January",1}
```
Out:
```
{1900, January, 1}
```

Note that although *Mathematica* has not displayed `January` in quotes, it is treated as a string; you can verify this by using the **FullForm** function to explicitly display the attributes of **myDate2**.

In:
```
FullForm[myDate2]
```
Out:
```
List[1900, "January", 1]
```

In each case, you can extract parts of the date without worrying about the type of data held in that part. Here, we use the double bracket ([[ ]]) form of the function **Part** which returns list parts.

In:
```
month=2;
myDate[[month]]
```
Out:
```
1
```

In:

> **myDate2[[month]]**

Out:

> January

Maintaining any structure inherent in your data has many advantages; related data can be kept in structures, and you can write object-oriented functions to operate on those structures (for example, see Maeder (1994) and Riddle & Dick (1994)).

1.4 Simple ASCII files

The simplest method of interchanging information between computer programs is to use ASCII files to store that information. The information components are easily read by eye, by word processor, or by spreadsheet and most applications programs support the import and export of data in this format. If you are authoring your own programs, writing information to a file in ASCII will be supported by the compiler or interpreter that you are using. For import into *Mathematica*, you might find it easier to always include a dividing character or space between, say, numbers. Tightly packed, formatted FORTRAN output, for example, will require a little more effort to read into *Mathematica* compared with the same output where, say, a space has been placed between every item output. Of course, number-separating spaces also make that information easier to read by eye.

In this section, we look at importing numbers and strings from files, but we also cover other nonexclusively related issues – string – number conversion, accessing variably named files, and content cataloging – that are of more general interest. So, even if you are disinterested in reading data from ASCII files for now, you still might want to browse through these other issues.

1.4.1 Numbers and strings from free-format ASCII files

Free-format files, in this context, are files in which each item in the file is bounded by a delimiting character, normally a space. The most straightforward function that you can use to read in free-format data is **ReadList**, which takes two arguments: the name of the file from which the data are to be read, and the type of data that are to be read. Here, we read in data from a file that contains a short list of metals and their melting points.

In:

> **myMetals=ReadList["Q127Alpha:Sam's observations:metals",**
> **{Word,Number}]**

Out:

```
{{Bismuth, 271}, {Lead, 327}, {Lithium, 179}, {Iron,
1537}, {Copper, 1085}}
```

There are several points worth noting. The filename is enclosed in double quotes and we have specified that we expect to read in one or more structures and that each structure consists of a word (the metal's name) and a number (the numeric melting-point of the metal). We have specified that each structure has a word and a number by enclosing those type names in curly brackets, *Mathematica*'s notation for a list. The types that *Mathematica* recognizes are: **Byte**, **Character**, **Real** (an approximate number in a FORTRAN-like format – for example, 1 or 1.342 or 1.3e4), **Number** (an exact number, such as 5 or 1025, or an approximate number in FORTRAN-like format), **Word** (delimited by word-separating characters that you can define), **Record** (delimited by record-separating characters that you can define), **String** (delimited by a new line), or **Expression** (a complete *Mathematica* expression).

Once these data have been read, we can access them individually or grouped in their structure by using the **Part** function. For example, the second part of the list **myMetals** can be obtained by

In:

```
Part[myMetals,2]
```

Out:

```
{Lead, 327}
```

or by the more usual form of the **Part** function, with double square brackets:

In:

```
myMetals[[2]]
```

Out:

```
{Lead, 327}
```

We reach the next level of the structure by specifying a further level with the **Part** function:

In:

```
myMetals[[2,2]]
```

Out:

```
327
```

By specifying **Number**, we have made *Mathematica* assume that an exact number is to be read. The file metalsAgain contains a mix of integer and real values. Because we have used **Number**, numbers read are left in their purest form: integers as integers, reals as reals.

In:

```
ReadList ["Q127Alpha:Sam's observations:metalsAgain",
         {Word,Number}]
```

Out:

```
{{Bismuth, 271.1}, {Lead, 327}, {Lithium, 179.3},
{Iron, 1537}, {Copper, 1085}}
```

Had we specified **Real** as the type, *Mathematica* would have converted all the numbers to their approximate form (identifiable by the omnipresence of the decimal point), regardless of the original form of the number:

In:

```
ReadList ["Q127Alpha:Sam's observations:metalsAgain",
         {Word,Real}]
```

Out:

```
{{Bismuth, 271.1}, {Lead, 327.}, {Lithium, 179.3},
{Iron, 1537.}, {Copper, 1085.}}
```

We also could have read in each line as a string by specifying the imported type to be **String**.

In:

```
myValues=ReadList ["Q127Alpha:Sam's observations:
                   metals",{String}]
```

Out:

```
{{Bismuth 271}, {}, {Lead 327}, {}, {Lithium 179}, {},
{Iron 1537}, {}, {Copper 1085}}
```

Note that there are several empty lists, **{}**, in **myValues**. These may be caused, for example, by blank lines in the file. In this instance, using **String** makes separating out the names and the melting points more difficult, should we want to do so later. In general, therefore, it is best to make all parts of your data as accessible as possible. For example, although we can still access each member of the list, it is now more difficult to access the numbers as separate entities – and an error message is generated if we use **Part** inappropriately.

In:

```
myValues [[3]]
```

Out:

```
{Lead 327}
```

In:

```
myValues[[3,2]]
```

Out:

```
Part::partw: Part 2 of {Lead 327} does not exist.
{{Bismuth 271}, {}, {Lead 327}, {}, {Lithium 179}, {},
{Iron 1537}, {}, {Copper 1085}}[[3,2]]
```

To overcome this problem, we can read each item within the file as a separate entity by using the type **Word**.

In:

```
myNewValues=ReadList["Q127Alpha:Sam's
                     observations:metals",
                     {Word}]
```

Out:

```
{{Bismuth}, {271}, {Lead}, {327}, {Lithium}, {179},
 {Iron}, {1537}, {Copper}, {1085}}
```

Note that when you read in data as type **Word**, each entity is read as a string. You will not be able to perform mathematical operations on the number-like strings directly because *Mathematica* treats strings as literals – and you need to display the full format of the string to see that it is indeed a string.

In:

```
myNewValues[[4]]+1
```

Out:

```
{1 + 327}
```

In:

```
FullForm[ myNewValues[[4]] ]
```

Out:

```
List["327"]
```

Of course, we can still access and mathematically manipulate the numbers, but to do so we need to know how to convert between strings and numbers.

1.4.2 String – number conversion

The ability to convert between numbers expressed as strings and numbers proper (expressed as numerals) enables you to extract usable numbers from strings (and vice versa) and to create variably valued strings that you can then use, for example, to access multiple files. We have seen that the elements of the list **myNewValues** are strings: when we display them fully using **FullForm**, the elements are enclosed in double quotes. If you want to manipulate a value encoded as a string, then you need to make the string a *Mathematica* expression. The function **ToExpression** converts its argument into a *Mathematica* expression; it will therefore convert a string to a manipulatable number. Here, we create the one-item-long list **myNumber**, whose (single) element we can then manipulate.

In:

```
myNumber=ToExpression[ myNewValues[[4]] ]
```

Out:

```
{327}
```

In:

```
myNumber+3
```

Out:

```
{330}
```

Here, we have converted a string to a number. You can also use **ToExpression** to convert any string (say, a symbolic formula) into a usable *Mathematica* expression.

The inverse of **ToExpression** is **ToString**. You can use the **ToString** function to make any *Mathematica* expression a string. You may find it useful to apply the function **FullForm** to see exactly how *Mathematica* is formatting your data; without using **FullForm**, the following three expressions – a list of numbers, a string version of that list, and a list of strings – appear identical:

In:

```
myNumbers=Table[i,{i,10}]//FullForm
```

Out:

```
List[1, 2, 3, 4, 5, 6, 7, 8, 9, 10]
```

In:

```
myString=ToString[myNumbers]//FullForm
```

Out:

```
"{1, 2, 3, 4, 5, 6, 7, 8, 9, 10}"
```

In:

```
myString2=Table[ToString[i],{i,10}]//FullForm
```

Out:

```
List["1", "2", "3", "4", "5", "6", "7", "8", "9", "10"]
```

1.4.3 Files with embedded comments

If your data files have comments, you may want to extract those comments for, say, a log of what each file contains or else to use the comment information to decide how to process the file. Most ASCII data files should contain a specific character at the beginning of each comment line to denote its comment status. We can use a "!" character for our comment lines, but any nonnumeric character will do.

In:

```
data=ReadList["Q127Beta:Alfy f:spreadCom.txt",String]
```

Out:

```
{! This is the spreadsheet file with comments,
!Time   V1   V2   I3,
1 0.841470985 0.666366745 0.560728281,
2 0.909297427 0.614300282 0.558581666,
3 0.141120008 0.990059086 0.139717146,
4 -0.756802495 0.727035131 -0.550222001,
5 -0.958924275 0.574400879 -0.550806946,
6 -0.279415498 0.961216805 -0.268578872,
7 0.656986599 0.791836209 0.520225778,
8 0.989358247 0.54922627 0.54338154,
9 0.412118485 0.916274317 0.377613584, }
```

To extract the comments, we just select out any lines beginning with the comment character. If your file-writing program is inconsistent about where it places the comment character – that is, if the special character is not always at the beginning of the line – you may have to check all the characters in a line for the comment character. Checking all the characters can take some time, but *Mathematica* only has to read the file once. One way to check a line's comment status is to test if its set of characters has the comment character as a member (using **MemberQ**), and then to select only those lines for which **MemberQ** returned **True**. Because **Select** takes as its second argument a function that returns **True** or **False**, we need to specify that **MemberQ[Characters[#],"!"]** is a function by appending *Mathematica*'s postfix function operator, **&**.

In:

```
Select[ data, MemberQ[Characters[#],"!"]&]
```

Out:

```
{! This is the spreadsheet file with comments,
!Time   V1   V2   I3}
```

Although **Select** works well to extract the comment lines, we want to minimize the number of passes through the file to select comments and numbers. After we have used the **MemberQ** function to detect if a line contains a comment, we can then use **Position** to sort out which lines contain numbers and which lines contain comments. We can then store the line numbers in pComs and pNums, respectively. Note that **Position** takes a pattern for its test, whereas **Select** takes a True-or-False test.

In:

```
coms = Map[MemberQ[Characters[#],"!"]&, data]
```

Out:

```
{True, True, False, False, False, False, False, False,
 False, False, False, False}
```

In:

```
pComs = Position[coms, True]
```

Out:

```
{{1}, {2}}
```

In:

```
pNums = Position[coms, False]
```

Out:

```
{{3}, {4}, {5}, {6}, {7}, {8}, {9}, {10}, {11}, {12}}
```

We can combine these operations into a function **ReadData** that we can reuse easily; we use **Print** to display the comment lines in a way that the user can easily cut and paste, but we leave the data as the answer returned by the function. We end the function with a simple filter **Select[vals, (Length[#] > 0)&]** to ignore blank lines; we only keep lines with lengths greater than zero.

In:

```
ReadData[path_String]:=
  Module[ {strs,coms,pComs,pNums,vals},
          strs = ReadList[path,String];
          coms = Map[MemberQ[Characters[#],"!"]&,
                     strs];
          pComs = Position[coms,True];
          pNums = Position[coms,False];
          coms = strs[[ Flatten[pComs] ]];
          Print /@ coms;
          strs = strs[[ Flatten[pNums] ]];
          vals=Map[ReadList[StringToStream[#],Number]&,
                   strs];
          Select[ vals, (Length[#]>0)&]
          ]
```

In:

```
data=ReadData["Q127Beta:Alfy f:spreadCom.txt"]
```

Print output:

```
! This is the spreadsheet file with comments
!Time   V1   V2   I3
```

Out:

```
{{1, 0.841470985, 0.666366745, 0.560728281},
 {2, 0.909297427, 0.614300282, 0.558581666},
 {3, 0.141120008, 0.990059086, 0.139717146},
 {4, -0.756802495, 0.727035131, -0.550222001},
 {5, -0.958924275, 0.574400879, -0.550806946},
 {6, -0.279415498, 0.961216805, -0.268578872},
 {7, 0.656986599, 0.791836209, 0.520225778},
 {8, 0.989358247, 0.54922627, 0.54338154},
 {9, 0.412118485, 0.916274317, 0.377613584}}
```

Our function **ReadData** can be improved. We need a help message so that anyone can find out what the function does, and a template for using the function. You also might want to include some options in the function, such as returning the comment lines with the data or not printing the comment lines. We describe the use of options in Chapter 2; they are also well explored by Maeder (1991).

In:

```
ReadData::usage ="ReadData[pathString] reads a file
with any number of columns. Blank and comment lines are
removed.";
```

1.4.4   Variably named files

The function **ToString** is useful when you need to access a group of files that may have been given some index number as part of their filenames. For example, a temperature-measuring device has been used to record temperatures inside three horticultural glasshouses, and the data have been stored in files with the glasshouse number as the last character of the filename. The data in these files could be read in one file at a time, but we can make *Mathematica* automatically read in the data from all the files. First, we create a table of filenames, each filename constructed by joining the non-variable part of the name to the indexing number, suitably converted from a number to a string. Next, we make a table of the three lists that were read in, one list per datafile, with each list containing four values.

In:

```
dataFilenames=
Table[
  StringJoin[
```

```
                    "Q127Alpha:Sam's observations:glasshouse.",
                    ToString[i]],
        {i,1,3}]
```

Out:

```
{Q127Alpha:Sam's observations:glasshouse.1,
Q127Alpha:Sam's observations:glasshouse.2,
Q127Alpha:Sam's observations:glasshouse.3}
```

In:

```
allData=Table[ReadList[dataFilenames[[i]],
                        Real],
              {i,1,3}]
```

Out:

```
{{12.4, 45.3, 23.8, 56.9}, {66.5, 68.7, 69.8, 72.3},
{-12.3, -14.3, -10.8, -9.5}}
```

You can easily extend this technique to cope with nonnumeric filenames. If the glasshouses were named, say, after the cardinal points of the compass, you could still use *Mathematica* to quicken the process of reading in your data:

In:

```
nameList={"north", "east", "west"}
```

Out:

```
{north, east, west}
```

In:

```
dataFilenames=
Table[
  StringJoin["Q127Alpha:Sam's observations:glasshouse.",
            nameList[[i]] ],
    {i,1,3}]
```

Out:

```
{Q127Alpha:Sam's observations:glasshouse.north,
Q127Alpha:Sam's observations:glasshouse.east,
Q127Alpha:Sam's observations:glasshouse.west}
```

In:

```
allData=Table[ReadList[dataFilenames[[i]],
                        Real],
              {i,1,3}]
```

Out:

```
{{12.4, 45.3, 23.8, 56.9}, {66.5, 68.7, 69.8, 72.3},
{-12.3, -14.3, -10.8, -9.5}}
```

Another variation on the theme of variably named files is the case where the next file to be read in is specified by the contents of the file currently open. For example, a data acquisition system produces a file containing numbers and, depending on the date of the experiment, stops inputting into the current file and starts inputting into another file. The name of the file that is about to be used is placed into the current file, as the last entry.

Until now, we have assumed that we knew the structure of a file – that is, what type of data was stored within the file and how the data were grouped. (We did not have to know how many elements of data there were; **ReadList** autonomously managed the end-of-file situation.) Now we have a new problem. Although we know that we want to read in numbers from the file, we do not know how many there are and, once the last number is encountered, we need to switch from reading numbers to reading the text containing the name of the next file to be read.

To cope with this new problem, we need a new set of tools; we cannot just use **ReadList** because at the end of the file our assumptions about the file's structure will break down. We need to be able to read in each element in the file, one at a time, and determine whether it is a number (in which case we add it to our list of numbers) or a filename.

Mathematica provides three functions that we can use to separate the tasks of opening, closing, and reading some element from a file. **OpenRead["f"]** opens file f for reading (and attaches that file to an I/O stream), **Read[s, t]** reads in one element of type t from an input stream s, and **Close[s]** terminates access to the stream s. (There is also the function **Skip[s, t]** that skips over an element of type t on input stream s.)

To tackle reading the contents of the file First experiment, we can read in each element as a **Word** type. After each element is read, we first test to make sure that we have not encountered the end of the file; *Mathematica* returns **EndOfFile** for any read operation that encounters the file end. Assuming we have not come to the file end, we convert the newly read element's string – formatted form into a number, using **ToExpression**. If the conversion succeeds (the function **NumberQ** returns **True**), the new number can be appended to the existing list. If the conversion fails, it must be part of the next file's name, and so it is inserted at the end of the **filenameString** variable, along with a single space to separate each word in the filename. Upon encountering the file end, we close that file and delete the spurious last space in the filename. (Although non-Macintosh computers would not have filenames with embedded spaces, this example could easily be extended to deal with other separating characters, if required.)

In:

```
(* initialize variables *)
   numberList={};
   filenameString="Q127Alpha:Sam's observations:";

(* open file *)
   inStream=
     OpenRead["Q127Alpha:Sam's observations:First
                experiment"];

(* process elements *)
   nextElement=Read[inStream,Word];
   While[TrueQ[nextElement=!=EndOfFile],

        If[NumberQ[ToExpression[nextElement]],
           AppendTo[numberList, nextElement],
           filenameString=StringInsert[filenameString,
                                       nextElement,
                                       -1];
           filenameString=StringInsert[filenameString,
                                       " ",
                                       -1];
          ];
        nextElement=Read[inStream,Word]
       ];
   Close[inStream];
   filenameString=StringDrop[filenameString,-1];
```

We can see the results by typing each of the variable names used.

In:

numberList

Out:

```
{12, 14, 17, 19, 16, 11, 9, 3, 0, -2, 0, 1, 0}
```

In:

filenameString//FullForm

Out:

```
"Q127Alpha:Sam's observations:Second experiment"
```

You can read a group of files with any names whatsoever if they are contained within a directory simply by using **SetDirectory** and **FileNames** to obtain a list of files within the directory. That list's members can then be used individually as the filenames for functions such as **OpenRead**.

1.4.5 File-content catalogs

Sometimes it is not possible to know what is in a file, but you might want to make a summary of the contents. One summarizing technique is to make a catalog of all the symbols found in a file and a count of the number of times each symbol occurs. To make such a catalog, we need an extensible list because we do not know *a priori* how many different symbols the file contains. For each symbol encountered in the file, we look through the list to check if the symbol is already present. If it is, we update its occurrence count; if it is absent, we add it to the list and set its occurrence count to one.

For example, here is a short program that lists the symbols that appeared in a paragraph of text, along with their occurrence count. First, it opens the text (ASCII) file called `That paragraph` which contains the text of the paragraph. Then, while the next character read from the file is not the end-of-file marker, it searches the list of previously found symbols to see if the new character is a member of that list. If the new character is a member, that character's occurrence count is incremented. If the new character is not a member, it is appended to the list of symbols found, with an occurrence count of unity. Lastly, the program closes the file, displays the list of found symbols, and exits.

In:

```
symbols={{"null",0}};
inStream=
 OpenRead["Q127Alpha:Sam's observations:That
         paragraph"];
nextElement=Read[inStream,Character];
While[TrueQ[nextElement=!=EndOfFile],
  If[MemberQ[Transpose[symbols][[1]],nextElement],
     symbols[[Position[Transpose[symbols][[1]],
              nextElement][[1,1]],2]]++,
    AppendTo[symbols, {nextElement,1}];
   ];
  nextElement=Read[inStream,Character];
];
Close[inStream];
symbols
```

Out:

```
{{null, 0}, {S, 1}, {o, 37}, {m, 15}, {e, 50}, {t, 43},
{i, 30}, {s, 28}, { , 104}, {n, 28}, {p, 3}, {b, 7},
{l, 18}, {k, 7}, {w, 7}, {h, 15}, {a, 27}, {f, 12}, {d,
11}, {y, 7}, {u, 16}, {g, 4}, {r, 10}, {c, 20}, {., 5},
{O, 1}, {q, 1}, {T, 1}, {,, 4}, {x, 1}, {F, 1}, {I, 1},
{;, 1}, {1, 1}}
```

To make **symbols** easier to assimilate, we can sort it into alphabetical order.

In:

```
Sort[%]
```

Out:

```
{{ , 104}, {,, 4}, {., 5}, {;, 1}, {a, 27}, {b, 7}, {c,
20}, {d, 11}, {e, 50}, {f, 12}, {F, 1}, {g, 4}, {h,
15}, {i, 30}, {I, 1}, {k, 7}, {l, 18}, {m, 15}, {n,
28}, {null, 0}, {o, 37}, {O, 1}, {p, 3}, {q, 1}, {r,
10}, {s, 28}, {S, 1}, {t, 43}, {T, 1}, {u, 16}, {w, 7},
{x, 1}, {y, 7}, {1, 1}}
```

Until now, we have worked in a procedural manner; each item read in has been processed when read. For many practical purposes, procedural working will "get the job done," and where future actions are dependent on the values being read, you may have no other choice. With *Mathematica*, however, you can chose to work in a functional style. For example, if you have a list of numbers (held in an array) that you want to sum, using a procedural language such as Pascal, then you would construct a small program to declare and populate the array and then add up each member of the array, one at a time. Your program might look like this:

```
CONST  howMany=12;
VAR myArray: ARRAY [1..howMany] OF INTEGER;
    i,sum:INTEGER;
    ...
    {fill the array}
    ...
sum:=0;
FOR i:=1 TO howMany DO
  BEGIN
    sum:=sum+myArray[i];
  END
```

You can program in *Mathematica* in such a procedural manner – but you can also tackle the same addition task by applying the function **+** (**Plus**) to the array.

In:

```
myArray={10,21,22,34,x};
sum=Apply[Plus,myArray]
```

Out:

```
87 + x
```

We hope that you recognize the greater efficiency of the functional method: no loop-counter or intermediate variables are required. Where possible, we shall adopt this functional style because it is more efficient than its procedural equivalent. It enables us to keep our work focused on the end application, without our becoming burdened with intermediate mechanisms. If you would like to know more about the different programming styles available within *Mathematica*, then the programming books by Maeder (1991, 1994) and Gaylord et al. (1993) are good reading.

Returning to our file-cataloging problem, if you want to know how many items of a particular kind are contained in a file, you can use the **Count** function to determine how many of the file's items match a given template. To count the number of a, T, and 1 characters there are in the file That paragraph, we adopt the following functional approach. First, we read in all the characters into a variable **allSymbols** and declare the list of characters whose occurrence count we want in **symbolSubset**. Second, we map the function **Count** over each of the members of **symbolSubset**. The function **Count** takes two arguments: the name of the list in which matches are to be found, and the template. **Count** is invoked three times, once for each of the members of **symbolSubset**; on each invocation, the **#** symbol is replaced by the next **symbolSubset** member. We have to wrap **Count** in **Function** so that *Mathematica* knows to interpret the symbol **#** as a dummy argument marker. Using **Map** in this way is equivalent to issuing the commands in sequence:

```
{Count[allSymbols,"a"], Count[allSymbols,"T"],
  Count[allSymbols,"1"]}
```

In:
```
inStream=
 OpenRead["Q127Alpha:Sam's observations:That
         paragraph"];
allSymbols=ReadList[inStream,Character];
Close[inStream];
symbolSubset={"a","T","1"};
Map[Function[Count[allSymbols,#]] , symbolSubset]
```
Out:
```
{27, 1, 1}
```

By wrapping **Function** around **Count**, we create a pure or anonymous function. Pure functions are important tools in *Mathematica*, and we devote a section in the Appendix to them. If pure functions are a new concept, you should consider reading that section now.

To catalog all the characters in **allSymbols**, we make an empty list and append to that list any symbol that is found in the file that is not already a member of that list. We can map the function "if it is not a member of list **s**, append it to **s**" over each of the symbols in **allSymbols**:

In:

```
s={};
Map[Function[If[!MemberQ[s, #],
                AppendTo[s, #]]],
    allSymbols];
s
```

Out:

```
{S, o, m, e, t, i, s,  , n, p, b, l, k, w, h, a, f, d,
 y, u, g, r, c, ., O, q, T, ,, x, F, I, ;, 1}
```

1.4.6   Numbers and strings from fixed-format ASCII files

In free-format files, we explained that each word or record is delineated by word- or record-separating characters. You can specify these characters by using the **WordSeparator** or **RecordSeparator** options in, say, **Read-List** or **FindList**. Data in fixed-format files cannot be accessed so easily. For example, using the **!!** function, we can see that the file Land area (in thousands of square miles, for the ten largest U.S. states (Economist, 1993)) contains a long list of ASCII text numbers:

In:

```
!!Q127Alpha:Sam's observations:Land area
```

Out:

```
5910266815871470121611401106104 1 978 971
```

Because the format of the data is fixed, there is no need to have any spaces between any of the items in the file. Although fixed-format files save a little disk storage space, they are not normally easy to read by eye! Indeed, without knowing the format in which the file was written, it is normally difficult to make any sense of its contents unless you are a part-time cryptanalyst. (The two gaps in the text do give a clue to its format, but this information alone would still not allow **ReadList** to be used to directly read numbers from the file because there are no separators between most of the numbers. **ReadList** would, by default, see only three numbers in the file.)

The file contains ten numbers and might have been written by the FORTRAN statements

```
        DIMENSION myArray(10)
        . . .
        WRITE(1,100)myArray
100     FORMAT(10I4)
```

or the C equivalent (ignoring the necessary file handling lines in both cases)

```
for(i=0; i<10; i++)
    fprintf(outStream, "%4d",myArray[i]);
```

Now that we know the format, we can read the numbers in blocks of n characters. One way to perform such a fixed-format read is to read in all the characters of the file as one-character strings, to convert each character to a one-digit number, to partition the resulting list into sublists of the correct length (four digits in this case), and then to convert the substrings into real numbers.

In:

```
inStream=
  OpenRead["Q127Alpha:Sam's observations:Land area"];
cList=ReadList[inStream,Character];
Close[inStream];
cList
```

Out:

```
{5, 9, 1, 0, 2, 6, 6, 8, 1, 5, 8, 7, 1, 4, 7, 0, 1, 2,
1, 6, 1, 1, 4, 0, 1, 1, 0, 6, 1, 0, 4, 1,  , 9, 7, 8,
 , 9, 7, 1, }
```

In:

```
cList[[1]]//FullForm
```

Out:

```
"5"
```

In:

```
mList=Map[Function[ToExpression[#]], cList]
```

Out:

```
{5, 9, 1, 0, 2, 6, 6, 8, 1, 5, 8, 7, 1, 4, 7, 0, 1, 2,
1, 6, 1, 1, 4, 0, 1, 1, 0, 6, 1, 0, 4, 1, Null, 9, 7,
8, Null, 9, 7, 1, Null}
```

In:

```
pList=Partition[mList,4] /. Null->0.
```

Out:

```
{{5, 9, 1, 0}, {2, 6, 6, 8}, {1, 5, 8, 7}, {1, 4, 7, 0},
{1, 2, 1, 6}, {1, 1, 4, 0}, {1, 1, 0, 6}, {1, 0, 4, 1},
{0., 9, 7, 8}, {0., 9, 7, 1}}
```

The conversion from sublist to number is a good example of **Map** and **Function** in use. The innermost **Map** multiplies the sublist members by the appropriate power of ten, as indicated by their positions in the sublist; the outermost **Map** adds together the sublist numbers and does so for each of the sublists in **pList**:

In:

```
Map[Function[Apply[Plus,
                   i=2;Map[Function[# 10^i--],#]]],
    pList]//N
```

Out:

```
{591., 266.8, 158.7, 147., 121.6, 114., 110.6, 104.1,
97.8, 97.1}
```

Had the data contained mixed fixed formats, then we could have used the **Take** function to extract certain members of the input stream for conversion specific to their format.

You should note that if the file-writing program changes the format of any of the numbers, for example, to add more precision, then any program (regardless of implementation language) attempting to read the file with the old format will produce erroneous results. Free-format data would still be read correctly – another good reason *not* to use fixed formats!

1.5   ASCII files created by spreadsheets

Data that have been exported from a spreadsheet into a text (ASCII) file are typically in the form of a matrix with a tab character separating numbers along a row and a carriage return (new line character) separating each row. In *Mathematica*, you can read in data stored in this format by treating each number as a word and each row as a record.

In:

```
ReadList[
  "Q127Alpha:Sam's observations:worksheet1 text format",
        Word, RecordLists->True,
        WordSeparators->{"\t"},
        RecordSeparators->{"\n"}]
```

Out:

```
{{1, 4, 7}, {2, 5, 8}, {3, 6, 9}, {10, 11, 12}}
```

We now discuss how to read in data stored in a more complicated spreadsheet format called DIF – the old VISICALC™ spreadsheet format. We go through DIF reading in some detail so that you can see how a number of techniques is used to tackle nontrivial formats. If you are not presently interested in such files, you might want to skip this section on a first read but to return to it later.

DIF files are still read and written by most spreadsheets and are used by some brands of test equipment. Later, we will look at DXF™ files, too. These are totally different from DIF files; DXF is the ASCII drawing exchange format used by the AutoCAD™ drafting program. The connection between these two different formats is that we can use **Position** to help read both DIF and DXF files. We look for keywords in ASCII files and, based on the position of the keywords, we can then extract combinations of characters, words, and numbers to effect the translation.

The spreadsheet file for the DIF translation is shown in Table 1:i.

Table 1:i   DIF-format file

```
Time V1 V2 I3
1 0.841470985 0.666366745 0.560728281
2 0.909297427 0.614300282 0.558581666
3 0.141120008 0.990059086 0.139717146
4 -0.756802495 0.727035131 -0.550222001
5 -0.958924275 0.574400879 -0.550806946
6 -0.279415498 0.961216805 -0.268578872
7 0.656986599 0.791836209 0.520225778
8 0.989358247 0.54922627 0.54338154
9 0.412118485 0.916274317 0.377613584
```

As before, we can read the file into a list of strings, one string per line. The DIF file was created by saving the spreadsheet file with the DIF format option active.

In:

```
strsDIF = ReadList["Q127Beta:Alfy f:spread.dif",
                String];
```

Files from different operating systems use different line termination characters. These line termination characters are usually some combination of linefeed (LF), or ASCII 10_{10}, and carriage return (CR), or ASCII 13_{10}. DOS-based Intel machines use both, as in CRLF. Apple Macintosh machines use only CR (some software on Power Macintosh machines uses LF), and Unix machines use only LF. These variations complicate our work because an operating system expects to see a certain line termination character(s), and any variation from what the operating system expects will cause extra characters to appear in the file, or may cause the file to appear as one long line. For example, if a DIF file was created on a CRLF-based operating system and is read on a Macintosh, the LF appears as an extra character. You may want to have a more flexible reader that looks for an extra (or missing!) ASCII character and deletes (or adds in) the appropriate character. Here we know the extra character appears on all lines but the first, so we will just delete the first character in each line after the first line.

To check for this, we print the first few lines as characters as shown below.

In:

```
tmp = Take[strsDIF, 4];
cTmp = Characters /@ tmp
```

Out:

```
{{T, A, B, L, E}, {, 0, ,, 1}, {, ", E, X, C, E, L, "},
{, V, E, C, T, O, R, S}}
```

Note how each line after the first begins with a comma but without an observable character. We can use the **ToCharacterCode** function to discover the unseen character. The following table contains in its top row the text from the second line, and in the second row the equivalent ASCII decimal character codes: the unseen character reveals itself as an LF. (This extra **LF** character arises because the DIF format is for DOS, but here we are reading the file on a Macintosh to illustrate potential problems.) Tables of ASCII characters can be written in decimal, hexadecimal, or octal, so be mindful of your number base when you determine which character is troublesome!

In:

```
TableForm[{cTmp[[2]],ToCharacterCode /@ cTmp[[2]]}]
```

Out:

```
0      ,      1
10     48     44     49
```

If we print about four lines of our file we can discover a few things about parsing a DIF file. First, the odd vertical offset for all lines after TABLE is due to the previous LF issue, which also makes the comma delimiters look more like apostrophes. Second, BOT (originally meaning Beginning of Tape) appears several times in the file; BOT is used as a row delimiter in DIF files. Finally, we detect groups of numbers separated by commas and by either text in quotation marks or the symbol **V**. The numbers separated by a comma denote a pair giving justification within a cell and cell numerical value. The cell numerical value is valid only if the next line contains the **V** symbol; when the next line contains a string, the string is printed with the given justification.

In:

```
Short[strsDIF,4]
```

Out:

```
{TABLE,    ,              ,             ,        ,     ,              ,       ,       ,
       0,1   "EXCEL"   VECTORS   0,10   ""   TUPLES   0,4   ""

      ,       ,      ,         ,        ,       ,              ,      , <<88>>,    ,
  DATA   0,0   ""   -1,0   BOT   1,0   "Time"   1,0                   V

           ,    ,                ,    ,              ,    ,
  0,0.41211849   V   0,0.91627432   V   0,0.37761358   V

        ,        }
  -1,0   EOD
```

So we will use the **Rest** function to keep *i*) every line after the first line and *ii*) every character after the first character on each of the kept lines. We use **StringJoin** to recombine the characters into a form that is easily searched by **Position**. The function **Position** is very useful for extracting information from files; in more complicated ASCII files, such as DXF files, **Position** would be used several times to find where different sections of information start and stop. BOT is used as a row delimiter in DIF files, so we find all of the positions of BOT in our file to mark out rows. Then we can analyze each row individually. The '**&**' symbol after **Rest[...]** is a shorthand notation for **Function[Rest[...]]**.

In:

```
tmp = Rest[strsDIF];
tmp = Map[Rest[Characters[#]]&, tmp];
tmp = Apply[ StringJoin, tmp, 1];
bots = Flatten[Position[tmp,"BOT"]]
```

Out:

```
{13, 23, 33, 43, 53, 63, 73, 83, 93, 103}
```

If we check our modified file with the **Short** function, we can see how removing the LF character has made some aspects of the file easier to read. Now no vertical offsets appear, but it is harder to see whether commas are a part of the information or are delimiters for the various lines.

In:

```
Short[tmp,4]
```

Out:

```
{0,1, "EXCEL", VECTORS, 0,10, "", TUPLES, 0,4, "", DATA,
 0,0, "", -1,0, BOT, 1,0, "Time", 1,0, "V1", <<87>>, V,
 0,0.41211849, V, 0,0.91627432, V, 0,0.37761358, V,
 -1,0, EOD}
```

By extracting the information between each occurrence of BOT, we get individual rows to parse, and so we preserve the tabular structure of the file.

In:

```
dt = Take[tmp, {bots[[1]]+1,bots[[2]]-1}]
```

Out:

```
{1,0, "Time", 1,0, "V1", 1,0, "V2", 1,0, "I3", -1,0}
```

Now we can find the column types in all of the even-numbered parts.

In:

```
cols = 2 Range[ Floor[Length[dt]/2.0] ]
```

Out:

```
{2, 4, 6, 8}
```

In:

```
colTypes = dt[[ cols ]]
```

Out:

```
{"Time", "V1", "V2", "I3"}
```

We could check these column types for either strings or the symbol **V**. If the columns are type **V**, then we would know to extract numerical information from the previous column, as we do below. If the column types are strings, then we would just store them as the information in the file.

In:

```
dt = Take[tmp, {bots[[2]]+1,bots[[3]]-1}]
```

Out:

```
{0,1, V, 0,0.84147098, V, 0,0.66636675, V, 0,0.56072828,
V, -1,0}
```

In:

```
cols = 2 Range[ Floor[Length[dt]/2.0] ];
colTypes = dt[[ cols ]]
```

Out:

```
{V, V, V, V}
```

Since these columns are type **V**, we extract the numbers from just after the commas in the previous line of the file, thus gathering the data.

In:

```
vals = Map[Drop[Characters[#],2]&,
           dt[[ cols-1 ]]];
Apply[ StringJoin, vals, 1]
```

Out:

```
{1, 0.84147098, 0.66636675, 0.56072828}
```

Now we can construct a function to read our simple file. We use mappings to variables such as **dt** and **cols** that represent lists of lists. Previously, **dt** and **cols** were simple lists containing the information from one row of data. Now, **dt** and **cols** will be lists of lists representing the entire spreadsheet. We need to exercise some care when parts of a row of **dt** are extracted. The part specification for **dt** needs two items separated by a comma: the first being the row number, the second being the column numbers desired from that row. The following example shows how we can turn **cols** and **dt** into matrices, and how the columns of an individual row must be extracted.

In:

```
rg = Range[Length[bots]-1];
dt = Map[Take[tmp, {bots[[#]]+1,bots[[#+1]]-1}]&, rg];
cols = Map[2 Range[ Floor[Length[#]/2.0] ]&, dt];
dt[[ 1, cols[[1]] ]]
```

Out:

```
{"Time", "V1", "V2", "I3"}
```

We need two coordinates, or specifiers, when extracting items from a matrix: the row number and the column number. We collect this coordinate information into the lists **cols** and **indx** and then collate them into a single matrix by using **Transpose**. (The function **MapIndexed** almost does what we want, but in the reverse order.) We then use **Apply** to insert the **cols** information with the full part specification into **dt**. We also need *i)* a function to extract the numerical values from the line before the column type and *ii)* a function to map **Characters** and **Drop** over this information. Since we are already applying the **If** statement over a list, we create and use the function **valExtract** to make the code more readable; a second pure function can make our code confusing to read.

In:

```
valExtract[data_] :=
    Apply[ StringJoin,
           Map[Drop[Characters[#],2]&,
               data],
           1]
```

In:

```
ReadDIF[ file_ ]:=
  Module[ {strsDIF, tmp, cols, dt, vals,
           colTypes, result, rg, indx, colsI, eod},
     strsDIF = ReadList[file,String];
     tmp = Rest[strsDIF];
     tmp = Map[Rest[Characters[#]]&, tmp];
     tmp = Apply[ StringJoin, tmp, 1];
```

```
        bots = Flatten[Position[tmp,"BOT"]];
        eod = Position[tmp,"EOD"][[1,1]];
        bots = Append[bots,eod];
        rg = Range[Length[bots]-1];
        dt = Map[Take[tmp, {bots[[#]]+1,bots[[#+1]]-1}]&,
                  rg];
        cols = Map[2 Range[ Floor[Length[#]/2.0] ]&,
                    dt];
        indx = Range[Length[cols]];
        colsI = Transpose[{indx,cols}];
        colTypes = Apply[dt[[##]]&, colsI, 1];
        Map[If[ colTypes[[#,1]] == "V",
                valExtract[ dt[[ #,cols[[#]]-1 ]] ],
                colTypes[[#]],
                colTypes[[#]]
              ]&,
            indx]
     ]
```

In:

```
    ReadDIF[ "Q127Beta:Alfy f:spread.dif" ] //TableForm
```

Out:

```
    "Time"    "V1"           "V2"           "I3"
        1     0.84147098     0.66636675     0.56072828
        2     0.90929743     0.61430028     0.55858167
        3     0.14112001     0.99005909     0.13971715
        4    -0.7568025      0.72703513    -0.550222
        5    -0.9589243      0.57440088    -0.5508069
        6    -0.2794155      0.9612168     -0.2685789
        7     0.6569866      0.79183621     0.52022578
        8     0.98935825     0.54922627     0.54338154
        9     0.41211849     0.91627432     0.37761358
```

(*Mathematica*'s formatting has limited the displayed precision of the numbers.) To make the function more robust, you should consider adding several checks and error messages. You should check to see that **strsDIF** is not empty to confirm that a nonempty file was read. You should also check that **bots** is not empty to confirm that **ReadDIF** handled the line termination characters correctly.

1.6 *Mathematica* files

Data written out to a file by the standard methods in *Mathematica* can be read in using the **Get** function; all directories specified in the system variable

$Path will be searched for the file. Note that you cannot use **Get** to read files that were not created by *Mathematica*.

In:

```
retrievedData=Get["myMathematicaData"]
```

Out:

```
{1, 2, 3.5, useful string, {{1, 2}, {3, 4}}}
```

Get retains any structures and types with which the data were stored.

1.7   DXF files

Sometimes ASCII file formats are a cross between a database and a language, such as the formats DIF or DXF. DXF is an ASCII drawing exchange database file format developed by AutoDesk, the makers of AutoCAD. The complete specification of DXF is in the AutoCAD program manual and is well beyond the scope of this book; you can find a summary in the collection of file formats by Kay & Levine (1992). In fact, our function **ShowDXF** will not display everything in any DXF file; however, it will display enough lines and text to cover most files. (We will describe enough of the workings of **ShowDXF** to give you an idea of how you can extend **ShowDXF**.)

The **ShowDXF** function we will develop uses many of the techniques discussed in the **ReadDIF** function. We will display only the simplest two-dimensional DXF files, but the **ShowDXF** can be easily extended to three-dimensional DXF files. We will not discuss all the details of every part of the **ShowDXF** function but instead will develop the main concepts and leave out most of the repetitive analysis of the different entity types.

A DXF file is made up of many short lines of text, as shown by **tmp**'s contents, below. Do not forget to make sure the file you are reading has the same carriage-return/linefeed line delimiter as your system, or otherwise you will get characters that will confuse **ShowDXF**.

In:

```
tmp = ReadList["Q127Beta:Alfy f:FILT.DXF",String];
TableForm[Take[tmp,15]]
```

Out:

```
0
SECTION
2
HEADER
9
```

```
$LTSCALE
40
1
9
$LIMMIN
10
0
20
0
9
```

The entities and blocks in the file are separated by a line code which uses a single zero, **0**, on a line. An entity is an object to be drawn. A block is one of more entities that have been grouped together and often repeated in many places on the drawing. **ShowDXF** is given below.

In:

```
ShowDXF::usage="ShowDXF[file,options] displays a DXF
file as a graphic.";
```

In:

```
ShowDXF[file_String, opts___Rule] :=
  Module[ {entities},
        entities = GetEntitiesDXF[file];
        DisplayEntitiesDXF[entities,opts]
  ]
```

The first part of the **ShowDXF** function reads the DXF file. The **GetEntities** function reads in the file, finds all of the group starting positions, and separates the file into groups. The second part of the **ShowDXF** function displays the entities by translating the DXF drawing commands into equivalent *Mathematica* drawing commands and then showing the results. Because some DXF files have several spaces before the zero, we use the first line of a DXF file as an example of the entity separator line. This makes the function more robust. We then convert as much of the file to *Mathematica* expressions as possible. The conversion creates a lot of error messages from **ToExpression**, which cannot match strings found in the file to known function or variable names, so in the final form of **GetEntitiesDXF** there are several **Off** commands to turn off these nugatory messages. The starting positions are converted into an array of numbers over which **Take** can be mapped to separate the individual groups.

In:

```
GetEntitiesDXF[file_String, opts___Rule] :=
  Module[ {tmp,dxf,dxfN,starts,groupPos,
            entities},
    dxf=ReadList[file,String];
    dxfN=Map[ToExpression, dxf];
    starts=Flatten[Position[ dxf, dxf[[1]] ]];
    groupPos=Transpose[{Drop[starts,-1]+1,
                        Drop[starts,1]-1}];
    entities = Map[Take[dxfN,#]&, groupPos]
  ]
```

The **DisplayEntitiesDXF** function is much longer than **GetEntities-DXF**. **DisplayEntitiesDXF** converts all the groups of graphic objects into objects that *Mathematica* can display. A DXF file also contains a lot of drawing setup information, such as colors and line pattern styles, which is not directly displayable but which does affect the rendering of the objects. For example, the first 460 lines of our FILT.DXF file contain drawing setup information.

Entities such as lines, arcs, and circles are directly translatable into graphics objects that *Mathematica* can show. We create a set of functions, **polyForm**, **lineForm**, **circleForm**, and **arcForm**, that are used to make *Mathematica* graphics objects from the DXF data. We generate these objects in response to occurrences of text specifiers such as LINE and CIRCLE, which we search for and whose positions we store – just as we did in the DIF reader. The **circleForm** function is shown below. The input list contains the basic information for the circle; the **xp**, **yp**, and **rot** variables are used to move the circle if it was defined as part of a block.

In:

```
circleForm[input_List,{xp_,yp_,rot_}] :=
  Module[ {x,y,r},
    x=input[[ Position[input,10][[1,1]]+1 ]];
    y=input[[ Position[input,20][[1,1]]+1 ]];
    r=input[[ Position[input,40][[1,1]]+1 ]];
    Circle[ move[{x,y},{xp,yp,rot}],r]
  ]
```

The flexibility of AutoCAD formats means more work for us when translating a DXF file. AutoCAD can define blocks of objects, and such blocks can be used many times in a file. Blocks can also be rotated and translated to any position. Thus part of our work involves finding blocks, determining what is in each block, and then replicating that block as many times as is specified, with any appropriate rotation and translation. The **move** function, used in the preceding **circleForm** function, uses the translation and rotation information (**xp**, **yp**, **rot**), as well as the original position information (x, y), to define the actual x-y position of the circle. The following code for **DisplayEntitiesDXF** shows how we can find blocks and turn them into graphics objects with the **buildGraphics** function, in the same way that simple graphics objects are handled. All of the objects are then joined into a long list and shown.

In:

```
DisplayEntitiesDXF[entityMatrix_List,opts___Rule] :=
  Module[{tmp,rowNames,secP,layNames,blockP,blockE,
          insP,mainP,lay,blockNames,gPrims,bPrims,
          insMatrix,ins,blockTakes,i},
(* convert linear file to matrix of entities *)
   lay = Layers /. {opts} /. Options[ShowDXF];
   If[ lay === All,
      (* pass it all *)
      entities = Select[entityMatrix,(Length[#]>0)&];
      rowNames = Apply[#1&, entities,1];
      layP = Flatten[Position[rowNames,LAYER]];
      layNames = layerName /@ entities[[layP]];
      Print["Layer Names = ", layNames];,
     (* select the requested layers *)
      entities = Select[entityMatrix,
      CheckLayer[#, lay]& ];
      rowNames = Apply[#1&, entities,1];
      Print["Layer Names = ", lay];
     ];
(* Select desired layers here or skip if All *)
   blockP = Flatten[Position[rowNames,BLOCK]];
   blockE = Flatten[Position[rowNames,ENDBLK]];
   blockTakes = Transpose[ {blockP+1,blockE-1}];
   blockNames = blockName /@ entities[[blockP]];
   Print["Block Names = ", blockNames];
   insP = Flatten[Position[rowNames,INSERT]];
(* find the block names to be inserted &
            their offset positions *)
   ins = insertForm /@ entities[[ insP ]];
   secP = Flatten[Position[rowNames,SECTION]];
   mainP = Last[secP];
(* build graphics from the main section *)
   gPrims = buildGraphics[Take[entities,
                            {mainP,Length[entities]}],
                       {0.0,0.0,0.0}];
(* build graphics from each block according to the move
   given by INSERT *)
```

```
    If[ Length[blockP]>0,
      insMatrix = insMove[ins,#]& /@ blockNames;
      bPrims = Flatten[
      Table[Map[buildGraphics[Take[entities,
                                    blockTakes[[i]] ], # ]&,
              insMatrix[[i]]],
            {i,Length[blockNames]}]
            ];,
      bPrims = {};
      ];
   Show[Graphics[Prepend[Join[gPrims,bPrims],
                        Thickness[0.0001]
                        ]
              ],
      AspectRatio->Automatic,opts]
   ]
```

The function **buildGraphics** does most of the real work: it searches the entity list for entities it recognizes, and it translates those found entities into *Mathematica* graphics commands.

In:

```
   buildGraphics[entities_List,offset_List] :=
     Module[ {tmp,rowNames,secP,layNames,blockP,insP,mainP,
              lineP,circP,arcP,textP,polyP,crs,ars,lns,
              blockNames,collectVs,gVs,ps},
       rowNames=Apply[#1&, entities, 1];
       lineP=Flatten[Position[rowNames,LINE]];
       circP=Flatten[Position[rowNames,CIRCLE]];
       arcP=Flatten[Position[rowNames,ARC]];
       textP=Flatten[Position[rowNames,TEXT]];
       polyP=Flatten[Position[rowNames,POLYLINE]];
       verP=Flatten[Position[rowNames,VERTEX]];
      (* Translate main graphics
         - those entities in Last section
        Forms should check for layers desired vs layers
          available *)
       If[ Length[circP] > 0,
          crs=Map[circleForm[#,offset]&,
                  entities[[ circP ]]];,
          crs={};
        ];
       If[ Length[arcP] > 0,
          ars=Map[arcForm[#,offset]&,
                  entities[[ arcP ]]];,
```

```
            ars={};
        ];
    If[ Length[lineP] > 0,
        lns=Map[lineForm[#,offset]&,
                    entities[[ lineP ]]];,
        lns={};
        ];
    If[ Length[verP] > 0,
        pC=Map[polyClosed, entities[[ polyP ]]];
      (* closed check *)
        gVs=groupVertex[ verP ];
        collectVs=Map[entities[[ # ]]&, gVs];
        collectVs=Table[Append[collectVs[[i]], pC[[i]]],
                    {i,Length[pC]}];
        ps=Map[polyForm[#,offset]&, collectVs];,
        ps={};
        ];
    Join[crs,ars,lns,ps]
    ]
```

A final concern is that the objects belong to layers in the drawing. For example, an engineer may use layers to hold objects, notes, or steps in a manufacturing process. We may want to display all or just a few layers, so we need first to select only those entities on layers we desire. The very first task of **DisplayEntitiesDXF** is to check the options for a list of layers or the word **All**. If a list of layers is given, then **Select** is used to compare each entities layer with that in the list and to keep the desired layers.

Not all of the code for **ShowDXF** has been repeated here. Our aim has been to show how various file types can be read, and to provide some useful code that you can develop further. All of the code for **ShowDXF** is given in the **DXFIN.MA** file. Here, we use **DXFIN.MA** to display, in *Mathematica*, a drawing of a T-configuration electrical filter that passes low frequencies.

In:

```
<<"Q127Beta:Alfy f:DXFIN.MA"
ShowDXF["Q127Beta:Alfy f:FILT.DXF"];
```

Out:

```
Layer Names = {1, 2, 3, 4, 5, 6, 7, 8, 9, 10}
Block Names = {}
```

The filter file uses polylines to describe a low-pass filter of two inductors and a capacitor. A side effect of **ShowDXF** is the listing of possible layers and found block names. Although a simple sketch was used here so that a short and easily understood example file could be built, we have used **ShowDXF** to display large electronic circuit schematics and other drawings.

1.8 Binary files

With data in ASCII files, it is possible to read in those data without much knowledge of the file format. Indeed, by using *Mathematica*'s word- and record-separator specifying facilities, considerable flexibility is available to you. With binary files this flexibility is reduced. The nature of a binary file is somewhat similar to fixed-format ASCII files: without knowing whether a sequence of bytes represents an integer or floating-point number with a specified format, it is nearly impossible to work out how you should read in the data.

In fact, even knowing whether a number has been written as an integer or as a floating point number is insufficient because number formats vary. Integers can be typically one, two, or four bytes long and may be stored with the least- or most-significant byte first. Floating-point numbers come in many formats. Both the IEEE-specified four- and eight-byte formats are the simplest for intermachine compatibility, but many compilers also produce their own formats which are optimized to use native floating-point hardware (for example, math co-processors). For example, the Symantec C compiler for the Apple Macintosh has four-byte `float` variables and eight-byte `short double` variables that correspond to the IEEE four- and eight-byte formats respectively, but it also has ten- and twelve-byte formats optimized for the Standard Apple Numerics Environment (SANE) and the MC68881 math co-processor, respectively. For the Power Macintosh, the same compiler supports only the native four- and eight-byte floating-point formats (both of which are IEEE-standard compliant.) Which format is in use is specified at compile time either by explicitly typing a variable (such as `float` or `short double`) or by switches within the compiler environment. Fortunately *Mathematica* is sufficiently flexible that most formats can be read.

The standard package `BinaryFiles.m` in the `Utilities` folder contains functions that will open a named binary file and that will read one-, two-, and four-byte integers (both unsigned and signed, the latter being assumed to be two's complement format) and four- and eight-byte IEEE-format floating-point numbers. There is also a function to read a C string (that is, a list of bytes containing the ASCII characters of a string and terminated by a zero-valued byte.) When you have finished reading data from a binary or ASCII file, you need to close that file using the **Close** function.

To illustrate handling input from binary files, we create and then read from a small binary file. The following C program (in ANSI C, except for the short double type which may be different on your compiler) opens a file called myDataFile, writes into the file four variables, each of a different type, and closes the file. If you are not familiar with C, you will want to know that the four arguments of the function **fwrite** are *i*) the address (specified by the **&** operator) of the variable to be written, *ii*) the size in bytes of the variable as supplied by the system function sizeof, *iii*) the number of items of that size to be written, and *iv*) the stream identifier.

```
#include<stdio.h>

main()
{

  FILE *myFile;

  short int myShort=2;
  long int myLong=876;
  float myFloat=12.43;
  short double myShortDouble=3.142e-12;
  unsigned long int itemsWritten;

/* open file; wb forces binary write access */

  myFile=fopen("myDataFile","wb");

/* write data */

  itemsWritten=fwrite(&myShort,sizeof(myShort),1,
                      myFile);
  itemsWritten=fwrite(&myLong,sizeof(myLong),1,myFile);
  itemsWritten=fwrite(&myFloat,sizeof(myFloat),1,
                      myFile);
  itemsWritten=fwrite(&myShortDouble,
                      sizeof(myShortDouble),1,myFile);

/* close file */

  fclose(myFile);

}
```

The *Mathematica* code required to read the contents of `myDataFile` into a list has only four lines, the first of which loads the package that handles binary files.

In:

```
Needs["Utilities`BinaryFiles`"];
myInputStream=
  OpenReadBinary[
      "Q127Alpha:Sam's observations:myDataFile"];
ReadBinary[myInputStream,
          {SignedInt16,SignedInt32,Single,Double}]
```

Out:

$$\{2,\ 876,\ 12.43,\ 3.142\ 10^{-12}\ \}$$

In:

```
Close[myInputStream];
```

Note that if you are reading files created on a different machine, you may need to specify the byte order. For example, a pointer to (that is, the address of) an integer on an Apple Macintosh points to the most significant byte; on an IBM PC, or a DEC VAX, it would point to the least significant byte. You can specify which byte order is in use with the **ByteOrder** option in the read function. Be aware that the only symptom of getting the byte order wrong will be incorrect numbers, and not an error message. (*Mathematica* cannot generate an error message because any combination of bits in an integer is a valid integer, even if the resulting number is of the wrong value.)

In:

```
myInputStream=OpenReadBinary[
    "Q127Alpha:Sam's observations:myDataFile"];
ReadBinary[myInputStream,
          {SignedInt16,SignedInt32,Single,Double},
          ByteOrder->LeastSignificantByteFirst]
```

Out:

$$\{512,\ 1812135936,\ 12.43,\ 3.142\ 10^{-12}\ \}$$

In:

```
Close[myInputStream];
```

1.9   References

Economist, "Pocket USA," Economist Books, London, UK, 1993.

Gaylord, R. J., Kamin, S. N., Wellin, P. R., "Introduction to Programming with *Mathematica*," Springer-Verlag, New York, USA, 1993.

Kay, D. C., Levine, J. R., "Graphics File Formats," McGraw-Hill, New York, USA, 1992.

Maeder, R., "Programming in *Mathematica*," Addison-Wesley, Redwood City, California, USA, 1991.

Maeder, R., "The *Mathematica* Programmer," Academic Press, Cambridge, Massachusetts, USA, 1994.

Riddle, A., Dick, S., "Applied Electronic Engineering with *Mathematica*," Addison-Wesley, Reading, Massachusetts, USA, 1995.

CHAPTER 2

Visualizing data

Most people find concepts and relationships between different quantities easier to grasp when the information is displayed visually. Our eyes and ears are excellent both at taking in large quantities of information, especially when all of it is presented to us in parallel, and at distilling that information into a small, concise set of notions in our consciousness. In this chapter, we show you how you can use *Mathematica* to help with data visualization – and, with *Mathematica*'s sound facilities, how to hear your data!

2.1   The art of visualization

The often-quoted phrase "a picture is worth a thousand words" is not necessarily correct. A graph that is drawn in an inappropriate way can cause confusion or even lead to a misinterpretation of the data it is illustrating.

For example, here is a graph of the number of days that the wind blew from a particular direction. The list **wd** contains pairs of numbers, each pair giving the azimuth and corresponding number of days. We use **ListPlot** to generate the graph.

In:
```
wd={{0,0},{80,5},{180,20},{225,15},{270,10}};
ListPlot[wd,
        AxesLabel->{"azimuth","days"},
        AxesOrigin->{0,0}]
```
Out:

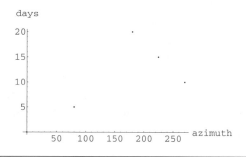

These data would be easier to comprehend if the graph looked more like the physical situation that it is representing. For example, you can show the polar nature of the data explicitly. (We show you how to make this graph later in the chapter.)

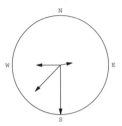

Mathematica has many built-in graphical functions, and most of the time you will be able to use one of these standard functions to display your data to best effect.

2.2 Standard graph-plotting functions

When you are looking for the right sort of plot for your data, there is no substitute for browsing through the sections "Graphics and Sound" and "The Structure of Graphics and Sound" in the *Mathematica* book by Wolfram (1991) and the "Guide to Standard *Mathematica* Packages," both supplied with *Mathematica*. Other relevant publications include "*Mathematica* Graphics" by Wickham-Jones (1994), "Applied *Mathematica*" by Shaw & Tigg (1994), and articles and columns appearing in *The* Mathematica *Journal*. This wealth of literature is sure to contain a graph that is at least close to what you want, even if it does not actually contain an already written graph-plotting function that you will be able to use.

Table 2:i gives a short directory that relates the type of data to one of *Mathematica*'s graph-drawing functions (where $y_i = f(x_i)$ for two-dimensional data or for apparently one-dimensional data, where the x_i are not given explicitly; $z_i = f(x_i, y_i)$ for three-dimensional data; and e_i is the error associated with y_i).

Be sure to explore the options for any function because they enable you to powerfully modify the basic plot style of their associated function. For example, in list plotting functions, **PlotJoined->True** will cause straight lines to be drawn between each point plotted; setting **PlotJoined->False** will cause only the points to be plotted.

You can list the options for any function by using **Options** [*function-Name*]. For information on a specific option, you use the normal *Mathematica* help request, that is, by placing a **?** in front of the function name on which help is sought.

Table 2:i   Data format and plotting functions

Data format	Possible functions
$\{y_1, y_2, y_3, y_4, y_5, \ldots\}$	ListPlot, BarChart, PieChart, TextListPlot, LabeledListPlot, LogListPlot, LogLogListPlot
$\{\{y_1, e_1\}, \{y_2, e_2\}, \ldots\}$	ErrorListPlot
$\{\{x_1, y_1\}, \{x_2, y_2\}, \ldots\}$	ListPlot, TextListPlot, LabeledListPlot
$\{\{x_1, y_1, z_1\}, \{x_2, y_2, z_2\}, \ldots\}$	ScatterPlot3D, TextListPlot, LabeledListPlot, ListSurfacePlot, ListShadowPlot3D
$\{\{z_{11}, z_{12}, z_{13}, \ldots\}, \{z_{21}, z_{22}, z_{23}, \ldots\}, \ldots\}$	ListContourPlot

In:

Options[ListPlot]

Out:

```
                           1
{AspectRatio -> -----------, Axes -> Automatic,
                GoldenRatio
AxesLabel -> None, AxesOrigin -> Automatic, AxesStyle
-> Automatic, Background -> Automatic, ColorOutput ->
Automatic, DefaultColor -> Automatic, Epilog -> {},
Frame -> False, FrameLabel -> None, FrameStyle ->
Automatic, FrameTicks -> Automatic, GridLines -> None,
PlotJoined -> False, PlotLabel -> None, PlotRange ->
Automatic, PlotRegion -> Automatic, PlotStyle ->
Automatic, Prolog -> {}, RotateLabel -> True, Ticks ->
Automatic, DefaultFont :> $DefaultFont, DisplayFunction
:> $DisplayFunction}
```

In:

?PlotJoined

Out:

```
PlotJoined is an option for ListPlot. With PlotJoined
-> True, the points plotted are joined by a line. With
PlotJoined -> False, they are not.
```

2.3   Augmenting standard *Mathematica* graphs

We started this chapter with a plot of wind directions: each arrow's direction showed the wind direction, and its length showed the number of days during which the wind had blown from that direction. The standard *Mathematica* package **Graphics`PlotField`** contains the function **ListPlotVectorField** that we used to generate the arrows. However, to make visualization easier, we placed a circle around the arrows and labeled it with the cardinal points of the compass. **Graphics`PlotField`** does not have a function for automatically encircling plots, so we added a labeled circle by defining a graphics object **circleBase**, which contains a circle and four items of text, the labels for the compass cardinal points.

Here is our *Mathematica* code for the wind survey plot. First, we load the **PlotField** package, and then we create the graphics object **circleBase** (that contains the labeled unit-radius circle) using two graphics primitives: **Circle** and **Text**.

In:

```
Needs["Graphics`PlotField`"];
circleBase=Graphics[{Circle[{0,0},1],
                     Text["N",{0,1.1}],
                     Text["W",{-1.1,0}],
                     Text["S",{0,-1.1}],
                     Text["E",{1.1,0}]}];
```

We then declare the experimental data, convert them to the format required for **ListPlotVectorField**, generate a plot filler (with suppressed graphical output because we have set **DisplayFunction->Identity**), and combine our background with the data using **Show**:

In:

```
windData={{0,0},{80,5},{180,20},{225,15},{270,10}};
windDataArrows=Map[{{0,0},
                    #[[2]] {Sin[#[[1]] Degree],
                            Cos[#[[1]] Degree]}/20}&,
                   windData] //N;
filler=ListPlotVectorField[windDataArrows,
                           DisplayFunction->Identity];
Show[circleBase,filler,
    PlotRange->All,
    DisplayFunction->$DisplayFunction,
    AspectRatio->Automatic];
```

Out:

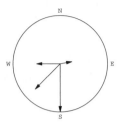

2.4 Data fusion

Data viewed in isolation often loses much of its significance. By adding a context, you can give the visualization much more impact and make it easier for your audience to see the full significance of your work. For example, an archaeological dig has resulted in several types of objects being found. Each object has its type and its {x, y} location carefully noted and stored in a list, **finds**.

In:

```
finds={{5,18,"bracelet"},{6,2,"jug"},{18,5,"pot"},
{4,6,"cup"}, {4,3,"cup"},{18,9,"pot"},{17,18,"pot"}};
```

A simple way of plotting this dataset is to use **TextListPlot** to write each object's type at its location.

In:

```
Needs["Graphics`Graphics`"];
TextListPlot[finds,PlotRange->{{0,21},{0,21}}];
```

Out:

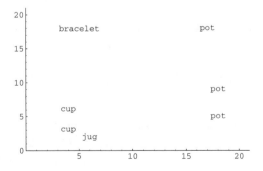

You can convert the text information to color to create a more eye-catching plot. Generating such a color plot takes a few steps. First, we have to draw markers (disks) at the position of each object. We need to be able to color each marker according to the type of object it is representing, too. *Mathematica* allows you color graphics primitives by placing the coloring directive in front of the primitive in a list. For example, when rendered, `{Hue[hVal], Disk[{x,y},r]}` will create a disk of radius **r** at the position **{x, y}** and will fill it with the hue **hVal**. Next, we need to create two graphics objects that contain *i)* the plot of the data and *ii)* the legend that relates the colors to the object type.

You can define the color of an object using red-green-blue, cyan-magenta-yellow-black, or hue (with optional saturation and brightness) descriptors. We use hue here because the **Hue** graphics directive can be used simply: its single argument cycles from 0 to 1, with the resulting color cycling from red through yellow, green, blue, and back to red. By writing a one-line function that takes in the string object type and returns a **Hue[n]** directive, we can color object markers according to their type. The numerical value assigned to any type is purely arbitrary; try changing the numbers to your own choice.

In:

```
myHue[z_String]:=Hue[Switch[z,
                    "bracelet",0.0,
                    "cup",0.15,
                    "jug",0.3,
                    "pot",0.6]];
```

We now map over the list **finds** a function that, for each object, gives a `{Hue[], Disk[]}` pair. The variable **dotGroundMap** defines a **Graphics** object of **finds**; no plot is generated because we have not applied the function **Show**.

In:

```
dots=Map[{myHue[#[[3]]],Disk[{#[[1]], #[[2]]}, 0.5]}&,
finds];
dotGroundMap=Graphics[dots,PlotRange->
{{1,21},{1,21}},Frame->True]
```

Next, we generate the legend graphics object.

In:

```
dotLegend=Graphics[{{myHue["bracelet"],
                    Disk[{0, 20}, 0.5]},
                    Text["bracelet",{5,20}],
                    {myHue["cup"],Disk[{0, 15}, 0.5]},
                    Text["cup",{5,15}],
                    {myHue["jug"],Disk[{0, 10}, 0.5]},
                    Text["jug",{5,10}],
```

```
            {myHue["pot"],Disk[{0, 5}, 0.5]},
            Text["pot",{5,5}]},
        PlotRange->{{0,21},{0,21}}]
```

Finally, we display the object map and its legend as an array of graphics objects (which looks much better in color!)

In:

```
    dotMap=Show[GraphicsArray[{dotGroundMap,dotLegend}]];
```

Out:

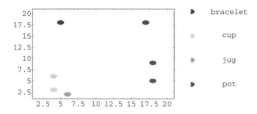

Although **dotMap** implies that there is some spatial clustering in the data, it conveys little other information. We know nothing about how the location of the objects correlates with, say, site topography.

Fortunately, one of the site archaeologists (or his graduate student) did make a height map of the site, taking the local height above some datum at each point in a 21-by-21 grid. These data have been stored in a file and have been read in to the variable **ground**. We can plot the height map as a density plot and then superimpose **dotGroundMap** on the height data to show that the found objects probably are correlated with terrain features.

In:

```
    groundMap=ListDensityPlot[ground];
```

Out:

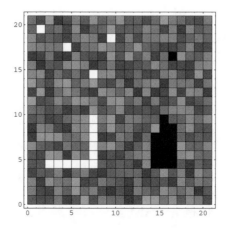

In:

```
Show[groundMap,dotGroundMap];
```

Out:

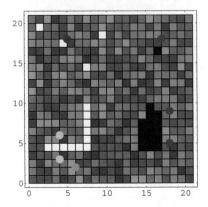

But a flat, two-dimensional plot of height is going against our principle of letting the plot show the nature of the data. A three-dimensional plot would be better.

Just as we used **Disk** to draw a filled circle on the two-dimensional site map, so we can use the three-dimensional graphics primitive **Point** (in the **'Graphics3D'** package, which we need to load) to place a colored point at a location **{x,y,z}**. Because we want the colored point to float above the ground (or else it would not be visible), we need to include the ground height, **z=ground[[x,y]]**, at each object's **{x,y}** position to give a location **{x,y,z}** at which each point can be drawn. To raise the point above the surface, we increase the point's **z** height by adding 2 to the ground height. Thus each object's data now becomes a list of *x*, *y*, *z*, and type.

In:

```
Needs["Graphics'Graphics3D'"];
finds3d=Map[{#[[1]],#[[2]],ground[[#[[1]],#[[2]]]],
#[[3]]}&, finds]
```

Out:

```
{{5, 18, 1.1, bracelet}, {6, 2, 1.05, jug}, {18, 5, 2.17,
pot}, {4, 6, 1.13, cup},{4, 3, 1.09, cup},{18, 9, 1.23,
pot}, {17, 18, 1.24, pot}}
```

The new list of primitives, **dots3D**, is colored using **Hue**, just as before. To use **ListSurfacePlot3D**, we need to make an array of {*x, y, height*} triplets, which we do using **Table**.

In:

```
dots3D=Map[{myHue[#[[4]]],
            Point[{#[[1]], #[[2]],2 + #[[3]]}]}&,
            finds3d]
```

Out:

```
{{Hue[0.], Point[{5, 18, 3.1}]}, {Hue[0.3], Point[{6,
2, 3.05}]}, {Hue[0.6], Point[{18, 5, 4.17}]},
{Hue[0.15], Point[{4, 6, 3.13}]}, {Hue[0.15], Point[{4,
3, 3.09}]}, {Hue[0.6], Point[{18, 9, 3.23}]},
{Hue[0.6], Point[{17, 18, 3.24}]}}}
```

In:

```
groundHeights=Table[{x,y,ground[[y,x]]},
                    {x,1,21},{y,1,21}];
surface=ListSurfacePlot3D[groundHeights,
            PlotRange->{{-2,23},{-2,23},{-1,5}},
            ViewPoint->{-1.212, -2.099, 2.361}];
```

Out:

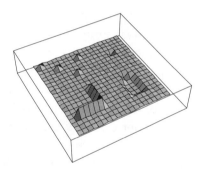

One final step produces a plot that shows object positions and the height information together.

In:

```
Show[surface,Graphics3D[{PointSize[0.05],dots3D}],
ViewPoint->{-1.2, -2.1, 2.4}];
```

Out:

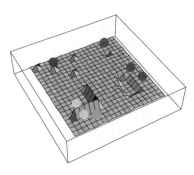

It is worth stating that by augmenting *Mathematica*'s basic graph functions with your custom application of graphics primitives, you create access to a graphics toolbox that is limited only by your imagination. As your requirements grow, you can take advantage of *Mathematica*'s extensibility to keep your data displays state-of-the-art.

2.5 Coping with awkward data

Not all data can be instantly used by one of *Mathematica*'s built-in functions, or even by a simple customization. In this section we show you some specific examples of how you can solve some unusual plotting requirements with quite general techniques.

2.5.1 Axis reversal

In some sciences, there is a requirement for axis reversal. For example, in astronomical data, the magnitude of an object decreases in numerical value as the object becomes brighter; a star with magnitude 3 is brighter than a magnitude 4 star. Such information could be plotted on a normal graph where smaller values are plotted lower down the y-axis, but then bright objects would be plotted below fainter objects – and astronomers like to plot brighter objects higher up the y-axis than fainter objects. Coordinates on the celestial sphere are another case where axis reversal is required. When a celestial object's Right Ascension or ecliptic longitude is plotted on the x-axis, the numeric values increase to the left so that any plot resembles closely how the objects would appear on the sky. (Right Ascension (RA) and Declination (Dec.) are to the celestial sphere what longitude and latitude are to the Earth.)

The simplest way to reverse the axis is to invert the sign of the y values and then to adjust the tick marks on the y-axis to read positive, despite the fact that they are plotted at negative values. The graphics option **Ticks->{{val, "text"},...}** allows us to label the tick mark at **val** with the given **text**. By specifying **Automatic** for the x-axis, we can keep the x-axis labels at their normal setting. For example, here we have seven observations of a star's magnitude, plotted with irradiance increasing up the y-axis.

In:

```
tMarks={{-11,"11"},{-10,"10"},{-9,"9"},
        {-8,"8"},{-7,"7"}};
mData={9.2, 8.6, 8.1, 8.2, 8.7, 9.2, 10.2};
```

```
ListPlot[-mData,
        Ticks->{Automatic,tMarks},
        PlotRange->{{0,Length[mData]},{-11,-7}},
        AxesLabel->{"observation","magnitude"},
        PlotJoined->True];
```

Out:

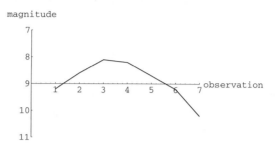

Our next example, a function that plots a map of Right Ascension and Declination, is somewhat more complicated. Although we preprocess the data and calculate values for some options, the plotting function that we use is just the basic **ListPlot**.

Here are the coordinates of some of the brighter stars in the equatorial constellation of Orion. We have expressed the Right Ascension and Declination values (in hours and minutes, and degrees and arcminutes, respectively) in a format where the decimal point separates the units; for example, 5.54 means 5 hours and 54 minutes. This method of expression is quite common, especially on pocket calculators.

In:

```
orion={{5.54,+7.24},
        {5.13,-8.13},
        {5.24,+6.19},
        {5.30,-0.18},
        {5.35,-1.13},
        {5.37,-2.37},
        {5.47,-9.40}};
```

Our function requires both coordinates to be expressed in normal decimal format, so we need to convert the list **orion** into this revised format by mapping a simple pure function onto the coordinate list.

In:

```
orionRADec=
  Map[{Quotient[#[[1]],1]+Mod[#[[1]],1]/.60,
      Sign[#[[2]]] (Quotient[Abs[#[[2]]],1]+
                    Mod[Abs[#[[2]]],1]/.60)}&,
      orion]
```

Out:

```
{{5.9, 7.4}, {5.21667, -8.21667}, {5.4, 6.31667}, {5.5,
-0.3}, {5.58333, -1.21667}, {5.61667, -2.61667},
{5.78333, -9.66667}}
```

Here is the plotting function. The star positions are passed in to the function as the list **p**; the second argument, **opts**, is a list of zero or more options, expressed as rules. Let us walk through the function. Because Right Ascension increases to the left, we make negative all the values of Right Ascension (storing them in **np**) so that we can use **ListPlot** without modification. We then override the default labels for the plot axes so that the labels are correct for the normal astronomical interpretation of the resulting map. So, most of the function is taken up with working out positions and text associated with the axes labels.

Taking control of tick mark and grid line placement makes it difficult for us to allow the user access to options that alter properties (like plot range, the origin of both axes, and where the grid lines are sited) that in turn impact on our code. However, we still want the user to be able to specify other options that do not affect our manipulation of the axes labels. Additionally, we want the user to be able to specify the range of Right Ascension and Declination that are to be plotted. We implement the specification of these ranges by the options **RARange->{**lower, upper**}** and **DecRange->{**lower, upper**}**.

The variable **keywords** holds a list of option names that are valid for **ListPlot**, but the cases **PlotRange**, **AxesOrigin**, and **GridLines** have been deleted from the full list. By selecting from **opts** only those options that are members of the abridged options list (that is, they are contained in **keywords**), we obtain a list of **ListPlot** filtered options, **lpfOpts**. That completes our tinkering with user-supplied options. Any option contained in **lpfOpts** is valid for **ListPlot** but is not one of those options over which we want control.

The next task is to calculate the range of the plot in x and y, so we work out those values by first extracting from the pairs of {*RA, Dec*} the appropriate parts (for example, only the first part, the Right Ascension, is required for the x-axis) and then finding the minimum or maximum by applying **Min** or **Max** to those lists.

The range of each axis is then calculated by using the minimum and maximum values just calculated – unless those values are overridden by an option supplied by the user. The x- and y-axis ranges are combined into one rule, **pRule**. The penultimate task is to calculate positions for the tickmarks and grid-lines, and their annotation values. Positions are straight-forward; annotations are complicated by our choice to express subdivisions of hours (in Right Ascension) and degrees (in Declination) in minutes.

In:

```
RADecPlot[p_List,opts___Rule]:=
  Module[{np,minX,maxX,minY,maxY,pRange,lpOpts,
          fOpts,myTicks,xTickInterval,
          yTickInterval,keywords,lpfOpts},

    np=Map[{-#[[1]],#[[2]]}&, p];

    keywords=Fold[DeleteCases,
                  Map[First, Options[ListPlot]],
                  {PlotRange, AxesOrigin, GridLines}];
    lpfOpts=Apply[Sequence,
                  Select[{opts},
                          MemberQ[keywords, First[#]]&]];

    minX=Apply[Min,Map[#[[1]]&,np]];
    minY=Apply[Min,Map[#[[2]]&,np]];
    maxX=Apply[Max,Map[#[[1]]&,np]];
    maxY=Apply[Max,Map[#[[2]]&,np]];

    xRange= -Reverse[RARange /.
                     {opts} /.
                     RARange->-{maxX,minX}];
    yRange= DecRange /.
            {opts} /.
            DecRange->{minY,maxY};

    pRule=PlotRange->{xRange,yRange};

    xTickInterval=4;
    yTickInterval=1/5;
    myTicks=Ticks->
      {Table[{xtVal,If[IntegerQ[xtVal],
                       FontForm[ToString[-xtVal],
                                {"Courier-Bold",14}],
                       FontForm[
                         ToString[-(xtVal-
                                     Ceiling[xtVal]) 60],
                                  {"Courier",9}]
                      ]},
              {xtVal,
               Floor[xRange[[1]]],
               Ceiling[xRange[[2]]],
               1/xTickInterval}],
```

```
            Table[{ytVal,If[IntegerQ[ytVal],
                        FontForm[ToString[ytVal],
                                {"Courier-Bold",14}],
                        FontForm[
                          ToString[(ytVal-If[ytVal>0,
                                            Floor[ytVal],
                                          Ceiling[ytVal]])
                                  60],
                              {"Courier",9}]
                              ]
                          },
                {ytVal,
                 Floor[yRange[[1]]],
                 Ceiling[yRange[[2]]],
                 1/yTickInterval}]
        };

    gridLines=GridLines->{Table[xtVal,
                              {xtVal,
                               Floor[xRange[[1]]],
                               Ceiling[xRange[[2]]],
                               1/xTickInterval}],
                          Table[ytVal,
                              {ytVal,
                               Floor[yRange[[1]]],
                               Ceiling[yRange[[2]]],
                               1/yTickInterval}]};

    axesOrigin=AxesOrigin->{xRange[[1]],yRange[[1]]};

    lpOpts=Sequence[pRule,myTicks,axesOrigin,lpfOpts,
    gridLines];
     ListPlot[np,lpOpts]
    ]
```

Finally, here is the map. Note how the standard options for **ListPlot** can still be given.

In:

```
RADecPlot[orionRADec,
        RARange->{4.5,6.5}, DecRange->{-15,15},
        AxesLabel->{"RA (hours)","Dec(degrees)"},
        PlotLabel->"Orion",
        Prolog->PointSize[0.05]]
```

Out:

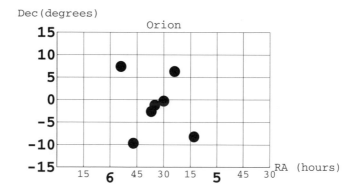

You can bring the tick mark and grid lines of the declination axis closer together by adjusting the value of **yTickInterval**. We suggest you try altering **RADecPlot** to accept, as options, different values of tick mark and grid-line separation.

2.5.2 Complex numbers

Mathematica provides the functions **Arg**, **Re**, **Im**, **Abs**, and **Conjugate** to help manipulate complex numbers. None of the standard plotting functions plots complex numbers directly, so you must split each complex number into its real and imaginary parts prior to plotting. For example, you can manipulate the list of complex numbers **cList** by writing a simple pure function that creates an **{x, y}** pair of real numbers for each member of **cList**.

In:

```
cList={1 + I, 2 + 0.5I, 3, 4 + -0.5I};
cxyList=Map[ {Re[#], Im[#]}&, cList]
```

Out:

```
{{1, 1}, {2, 0.5}, {3, 0}, {4, -0.5}}
```

Alternatively, you can express complex numbers in polar form {*r, theta* (radians)}. Note how *Mathematica* has preserved the integer exactness of the first complex number. (You can use the enumeration function **N** to convert the list to real numbers.)

In:

```
cpolarList=Map[ {Abs[#], Arg[#]}&, cList]
```

Out:

$$\{\{Sqrt[2], \ -\frac{Pi}{4}\}, \ \{2.06155, \ 0.244979\}, \ \{3, \ 0\}, \ \{4.03113,$$

$$-0.124355\}\}$$

In:

cpolarList=Map[{Abs[#], Arg[#]}&, cList] //N

Out:

```
{{1.41421, 0.785398}, {2.06155, 0.244979}, {3., 0},
{4.03113, -0.124355}}
```

Once you have your data in {*x, y*} form, then you can use any of the normal pair-plotting functions.

2.5.3   Unwrapping cyclic phase data

Sometimes data are stored in a way that makes analysis awkward. Electronic network analyzers store phase data in a sawtooth pattern that folds over from −180° to +180° and makes it difficult to answer the two main questions associated with a network's phase response: "As the incoming signal's frequency increases, what is the network's phase delay?," and "what is the network's phase ripple?" The following *Mathematica* code creates 201 points of data that suffer from phase wrap-around, with a small sine wave superimposed to act as ripple.

In:

```
phase = -Table[11.3 i + Sin[0.1 i],{i,0,200}] //N;
inc = 0;
pwrap = Table[ If[(phase[[i]]+inc 360) < -180.0,
                  inc++
                  ];
              phase[[i]]+inc 360,
            {i,Length[phase]}
            ];
ListPlot[pwrap,PlotJoined->True]
```

Out:

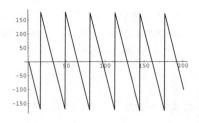

We used **Table** in a relatively straightforward way to create the phase data, but our unwrapping code will work on entire arrays to make it more efficient. The basic idea is to find all of the discontinuities in our data by calculating a first-order difference, $x_i - x_{i-1}$. At each discontinuity found, we accumulate a 360° phase shift which we then subtract from our data. We round down any differences less than 360° to zero.

In:

```
diff = Drop[pwrap,1]-Drop[pwrap,-1];
accum = FoldList[Plus,0,360 Round[diff/360]];
Unwrap[x_List]:=Module[{diff},
                    diff = Drop[x,1]-Drop[x,-1];
                    x - FoldList[Plus,0,
                                360 Round[diff/360]]
                    ]
```

In:

```
ListPlot[Unwrap[pwrap],PlotJoined->True]
```

Out:

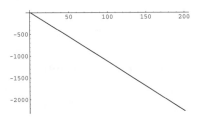

Once we have the unwrapped data, we can use **Fit** to find the delay.

In:

```
linFit = Fit[Unwrap[pwrap],{1,x},x]
```

Out:

```
11.0788 - 11.2981 x
```

In:

```
delay = -Coefficient[linFit,x]
```

Out:

```
11.2981
```

In:

```
ListPlot[Unwrap[pwrap] + delay Range[Length[pwrap]],
        PlotJoined->True]
```

Out:

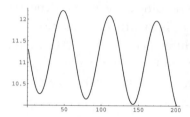

Once the linear delay from **Fit** is added to the unwrapped data, we are left with the phase ripple. The delay is fit to 11.298, which is very close to the 11.3 value used to create the data. The ripple plot shows a peak ripple of 1.0, as was used in the original data, and a small residual phase slope due to the difference between 11.3 and 11.298.

2.6   Making your own graphs

If you decide that the best way to visualize your data is to design from scratch your own graph plotting function, *Mathematica* provides you with many graphics primitives that you can use as building blocks. In this section, we describe some build-your-own graph-making functions. But first you need to know what graphics primitives are available, because these are the building blocks with which you have to work.

The main set of primitive objects includes **Point**, **Line**, **Rectangle**, **Circle**, **Disk**, **Cuboid**, and **Text**. You can attach to these objects such attributes (where appropriate) as **GrayLevel**, **Hue**, **RGBColor**, **CMYKColor**, **Dashing**, **Thickness**, **AbsoluteThickness**, **PointSize**, **FontForm**, and **AbsolutePointSize**.

Many of the primitive functions use function overloading (as in C++) to provide a variety of end results from one object type. For example, text can be drawn at a location specified by cartesian coordinates:

In:
```
Show[Graphics[Text["string",{1,1}],Frame->True]]
```
Out:

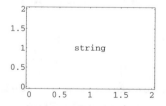

The overloaded form of **Text** enables more attributes to be specified; the *Mathematica* book by Wolfram (1991) describes the various arguments that **Text** can take. Here, using two more arguments to the **Text** function, we specify not only the text and its location but also how it is offset and laid out:

In:

```
location={1,1};
centerAbove={0,-1};
topToBottom={0,-1};
offset={10,0};
Show[
  Graphics[Text[FontForm["string",{"Times",9}],
              location,
              centerAbove,
              topToBottom],
          Frame->True]]
```

Out:

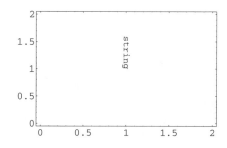

2.6.1 Polar surface plot

Our first example of make-your-own graphics code shows how you can write a polar surface map. One use of such a map is in biology, where you might want to show mold growth density on a gel in a Petri dish. Although we could use the standard surface plotting routines, these tend to be rectangle-oriented. A better approach is to maintain the shape of the round Petri dish in the plot – hence our polar surface map.

Before we begin work on the plotting function, we need to know how the data are organized. For this example, we use data that are divided into annular rings, each ring having a fixed cross section. In turn, each annulus is divided into sectors. Although we assume any annulus has sectors of a fixed angular size, the annuli can have different numbers of sectors. For example, it might be sensible to divide the Petri dish area into three annuli with the inner, middle, and outer annuli having one, four, and eight sectors each, respectively.

We specify the data in the form of a list of lists: the first inner list contains the data for the outermost annulus, the next contains the data for the adjacent annulus, and so on. The number of inner lists defines the number of annuli; the number of elements of each inner list defines the number of sectors in the corresponding annulus.

In:

```
petri={{5,7,6,4,4,6,5,6},{8,9,7,9},{12}};
```

This section contains the function **ListPolarMap** and the help messages for the function. The function expects two arguments and zero or more options.

The first argument is the name of a list that contains m lists of n values (for example, **petri**). The second argument is the name of a function that we defined and that translates numeric values into hues or grayscale shades.

We define four options:

- **GridLines->**x where x is either **True** or **False**; if **True** then black lines are drawn on the plot to delineate the boundaries of the annuli and their sectors, creating a spider's web effect.
- **PlotName->"**$string$**"** gives the plot a title; a new line can be forced by inserting **n** in the string.
- **Legend->{{**$val1$,"$string1$" **},{**$val2$,"$string2$"**} ,.......}** creates a legend for the colors. The legend is placed to the right of the plot. We have not created an automatic legend.
- **EastToRight->**x where x is **True** or **False**. We have labeled the plot with the cardinal points of the compass, and this option allows you to swap the East-West sense of the plot, allowing the plot to represent a from-above or a from-below image.

For any function, but especially for large functions, it is useful to have a help message so that the user can inquire about how to call the function and what options it takes. We also create messages that are printed out when the function detects an error.

In:

```
ListPolarMap::usage="ListPolarMap[d,f] creates a \
polar colour map of the data in d using the colouring \
function f. d is a list of annular value lists, \
with each successive list containing values in \
different sectors in each annulus(outermost annulus \
first).";


Options[ListPolarMap]={GridLines->True,
                        EastToRight->True,
```

```
                        PlotName->"",
                        Legend->{{}}
                };
```

```
ListPolarMap::unknGrOp="ListPolarMap called \
with an unknown option for GridLines. \
GridLines->True or False.";
```

```
ListPolarMap::unknEWOp="ListPolarMap called \
with an unknown option for EastToRight. \
EastToRight->True or False.";
```

The function operates as follows. We declare local variables, and we set the values of the options either to those specified by the user or else to their default values. The plot itself is built up in the variable **plotList**. First, we draw the colored sectors for all the annuli, then (if specified) a legend and grid lines, and finally the plot title and cardinal point labels.

In:

```
ListPolarMap[data_,colorFunc_,options___Rule]:=
  Module[
    {numAnnuli,a,r,plotList,
     legendBoxSize=0.15,
     gridLinesWanted,name,gridList,
     legendList,lPlotList,
     plotCW,legendXOffset=1.75,
     nameYOffset=1.5,systemFont},

    systemFont=$DefaultFont;

  (* evaluate options *)

    gridLinesWanted=GridLines /. {options} /.
                    Options[ListPolarMap];
    name=PlotName /. {options} /.
         Options[ListPolarMap];
    legendList=Legend /. {options} /.
              Options[ListPolarMap];
    plotCW=If[EastToRight /. {options} /.
              Options[ListPolarMap],
            -1,
            +1,
            Message[ListPolarMap::unknEWOp];
          ];
```

```
        numAnnuli=Length[data];

    (* generate a list of {colour, disk_description}
    pairs; color of disk is determined by result of
    function 'colorFunc', the disk is defined by arguments
    Disk[{x0,y0},r,{theta0,theta1}]. The actual generation
    of the list is by calling Table to make a
    two-dimensional table which is then reduced in
    dimension to a one-dimensional list using Flatten. *)

        plotList=
        Flatten[
           Table[{colorFunc[data[[r,a]]],
                 Disk[{0,0},
                     1-(r-1)/numAnnuli,
                        If[plotCW==-1,
                        {Pi/2 + plotCW a 2 Pi /
                                    Length[data[[r]]],
                         Pi/2 + plotCW (a-1) 2 Pi /
                                    Length[data[[r]]]},
                        {Pi/2 + plotCW (a-1) 2 Pi /
                                    Length[data[[r]]],
                         Pi/2 + plotCW a 2 Pi /
                                    Length[data[[r]]]}
                        ]
                     ]
                 },

                 {r,1,numAnnuli},
                 {a,1,Length[data[[r]]]}
                 ]
           ,1];

    (* generate a list of coloured rectangles for
       the legend *)

    If[Length[legendList[[1]]]>0,

        lPlotList=
           Table[{{colorFunc[legendList[[i,1]]],
                  Rectangle[{legendXOffset-
                     legendBoxSize,
                     1.0-legendBoxSize*i},
                     {legendXOffset,
                     1.0-legendBoxSize*(i-1)}]},
```

```
                    Text[legendList[[i,2]],
                        {legendXOffset+0.1,
                         1+legendBoxSize/2-
                             legendBoxSize*i},
                        {-1,0}]
                },
                {i,1,Length[legendList]}
                ];
        AppendTo[plotList,lPlotList];
    ];


(* generate the grid lines *)


    Switch[gridLinesWanted,
        True,gridList=
            Flatten[
                Table[{GrayLevel[0],
                        Circle[{0,0},
                                1-(r-1)/numAnnuli]},
                {r,1,numAnnuli}],1];
                AppendTo[plotList,gridList],
        False,,
        _,Message[ListPolarMap::unknGrOp];
            Return[]];

    For[r=1,r<=numAnnuli,r++,
      If[Length[data[[r]]]!=1,
      gridList=Flatten[
          Table[{GrayLevel[0],
                Line[{(1-r/numAnnuli)*
                      {Sin[(a-1) 2 Pi /
                          Length[data[[r]]]],
                       Cos[(a-1) 2 Pi /
                          Length[data[[r]]]]},
                      (1-(r-1)/numAnnuli)*
                      {Sin[(a-1) 2 Pi /
                          Length[data[[r]]]],
                       Cos[(a-1) 2 Pi /
                          Length[data[[r]]]]}}]},
                {a,1,Length[data[[r]]]}],1];
      If[gridLinesWanted==True,
        AppendTo[plotList,gridList]];];
    ];
```

```
            (* draw plot name *)

          If[name!="",
             PrependTo[plotList,
                   Text[FontForm[name,{"Times-Bold",14}],
                           {0,nameYOffset},{0,1}]]];

       (* label polar plot *)

          $DefaultFont={"Courier-Oblique",10};

          PrependTo[plotList,Text["North",
                                    {0,1.05},{0,-1}]];
          PrependTo[plotList,Text["South",
                                    {0,-1.05},{0,1}]];

          If[plotCW==-1,
             PrependTo[plotList,Text["West",
                                    {-1.05,0},{1,0}]];
             PrependTo[plotList,Text["East",
                                    {1.05,0},{-1,0}]]; ,

             PrependTo[plotList,Text["West",
                                    {1.05,0},{-1,0}]];
             PrependTo[plotList,Text["East",
                                    {-1.05,0},{1,0}]];
             ];

          $DefaultFont=systemFont;

          Show[
             Graphics[
                {Rectangle[{-6,-4},{6,4},
                   Graphics[plotList,
                            PlotRange->All,
                            AspectRatio->Automatic
                           ]
                          ]
                },
                AspectRatio->Automatic,
                PlotRange->All
               ]
             ] (* end Show *)

       ] (* end Module *)
```

The second argument to **ListPolarMap** is the coloring function for the data. Here are two coloring functions: a graylevel generator and a hue generator. The chosen color function is used by the **Disk** and **Rectangle** primitives. You will find it useful to experiment with both functions so that the resulting plot shows your data appropriately.

In:

```
myGL[x_]:=GrayLevel[0.1+x/14];
myHue[x_]:=Hue[0.1+x/14];
aa=ListPolarMap[petri,myHue,GridLines->True,
PlotName->"culture A (from above)",
Legend->{{0.0,"zero"},{3,"3"},{6.,"6"},
        {9.,"9"},{12.,"12"}}];
```

Out:

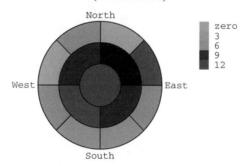

To draw the graylevel version, we would use

In:

```
aa=ListPolarMap[petri,myGL,GridLines->False,
PlotName->"culture A (from below)", EastToRight->False,
Legend->{{0.0,"zero"},{3,"3"},{6.,"6"},
        {9.,"9"},{12.,"12"}}];
```

We suggest you try altering both the instances of **colorFunc** within **List-PolarMap** and the definition of your chosen color function to take a second argument – for example, by passing the data's maximum value to the color function, you can make it automatically scale the color.

In:

```
(* code extract *)
.
colorFunc[val, Max[dataList]]
.
.
myHue[x_,max_]:=Hue[0.1+x/max];
```

2.6.2   Bar charts and Manhattan plots

Bar charts are very useful for observing the statistics of data distributions, and when data have more than one parameter, then their bar charts become multidimensional. *Mathematica* contains standard functions, such as **BinCount**, **BarChart**, and **BarChart3D**, for counting distributions and plotting one- and two-dimensional distributions. In this section we show you how to create some functions for nonstandard cases. We will create one- and two-dimensional binning functions for nonuniform bins. Then we will create a plotting function for displaying the nonuniformly spaced one- and two-dimensional bins. Nonuniform bins can arise from manufacturing specification lists or from physical constraints, such as counting items in predetermined land areas.

Let us examine the following data with uniform and nonuniform bin spacing. We use **Short** to give us a summary.

In:

```
Short[data1 = Table[Random[],{100}] ]
```

Out:

```
{0.357169, 0.712127, 0.84456, <<95>>, 0.267954,
 0.090515}
```

In:

```
Needs["Statistics`DataManipulation`"]
?BinCounts
```

Out:

```
BinCounts[data, {xmin, xmax, dx}] gives a list of the
number of elements in data that lie in bins from xmin
to xmax in steps of dx.
   BinCounts[{{x1,y1}, {x2,y2}, ...}, {xmin, xmax, dx},
{ymin, ymax, dy}] gives an array of bin counts.
```

In:

```
b1 = BinCounts[ data1, {0,1,0.1} ]
```

Out:

```
{8, 9, 14, 11, 11, 10, 9, 13, 11, 4}
```

In:

```
Needs["Graphics`Graphics`"]
BarChart[ b1 ];
```

Out:

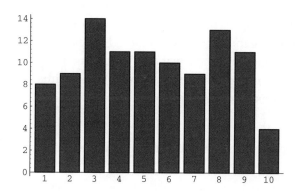

When our bins are nonuniformly spaced, we can use **Select** to gather together the data that fall in a particular bin. Possibly the most straightforward way of counting the data would be to use **Select** to cut from the list the items above and below each bin's specification. Then we can use **Length** to count the number of items gathered and **Table** to step through all the specifications while gathering the results into a list. Our function **histo1** works in this way.

In:

```
histo1[data_List,spec_List] :=
  Module[{i},
    Table[Length[Select[Select[data,(#<=spec[[i+1]])&],
        (# > spec[[i]])& ]], {i,Length[spec]-1}]
  ]
```

By using a set of bin widths which is wider in the center, we should give the distribution a more rounded appearance. Arbitrarily, we will use a cosine curve to specify the spacing of the bins.

In:

```
spec = 0.5(Cos[ Pi Range[0,1,0.1] + Pi] + 1.0) //N
```

Out:

```
{0., 0.0244717, 0.0954915, 0.206107, 0.345492, 0.5,
0.654508, 0.793893, 0.904508, 0.975528, 1.}
```

In:

```
bc = histo1[ data1, spec]
```

Out:

```
{4, 4, 10, 18, 17, 15, 16, 12, 4, 0}
```

(As an aside, computational speed is often an issue when analyzing statistics. We have chosen some very small data sets here, but often the data sets are large. Using **Table**, **Do**, and **For** can often result in code that is easily programmed and readily understood by ex-FORTRAN programmers, but that is inefficient in *Mathematica*. Processing arrays with **Map** or **Apply** usually results in faster *Mathematica* code. We can time the **histo1** and **BinCounts** functions to show how long it takes to analyze 100 points with 10 bins.

In:

```
Timing[ histo1[ data1, spec] ]
```

Out:

```
{0.616667 Second, {4, 4, 10, 18, 17, 15, 16, 12, 4, 0}}
```

In:

```
Timing[ BinCounts[ data1, {0,1,0.1} ] ]
```

Out:

```
{0.05 Second, {8, 9, 14, 11, 11, 10, 9, 13, 11, 4}}
```

The times show **BinCounts** to be much faster. But remember that the results are not the same because we used different bin widths for **histo1** and for **BinCounts**.)

We can speed up our histogram function by mapping **Select** over the data rather than embedding it in **Table**. Our new histogram code involves two auxiliary functions, **selectLess** and **selectGreater**, so that the pure functions needed by **Select** and **Map** do not conflict. **selectLess** creates a set of data lists. Each list is longer than the previous one, and the number of lists is equal to the number of bins. Next, we group the lists with the maximum bin specification and use **selectGreater** to trim the lists into bins. An auxiliary result of this function could be the binned data, such as those given by the **BinLists** function in the **DataManipulation** package. However, for computing just the histogram, only the length of each bin is returned.

In:

```
selectLess[d_,v_] := Select[ d, (#<=v)&]
selectGreater[d_,v_] := Select[ d, (#>v)&]
histo1m[data_List,spec_List] :=
  Module[ {dt,lis},
    dt = selectLess[ data, #]& /@ Drop[spec,1];
    lis = Transpose[{dt,Drop[spec,-1]}];
    Length /@ Apply[ selectGreater, lis, 1]
  ]
```

This function is almost three times as fast as the original function for this short data list, and the relative speed will increase as the data list gets longer. Both of these functions are slow compared to **BinCount**, which first converts the data to integers and then uses **Count**.

In:

```
Timing[ bc=histo1m[ data1, spec] ]
```

Out:

```
{0.216667 Second, {4, 4, 10, 18, 17, 15, 16, 12, 4, 0}}
```

In:

```
BarChart[bc];
```

Out:

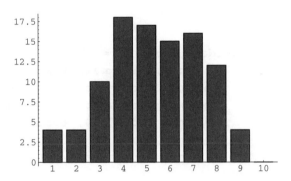

Although the distribution reflects the changes in bin widths, the bar chart does not show the changes in the specifications. Because the graphics primitives in *Mathematica* are very powerful and easy to use, we can construct our own bar chart, called **BoxPlot**, using rectangles and standard axes. From the help message, we see that each rectangle needs two points for its definition. We can use the data and bin-limits specification lists to provide these points.

In:

```
?Rectangle
```

Out:

```
Rectangle[{xmin, ymin}, {xmax, ymax}] is a
two-dimensional graphics primitive that represents a
filled rectangle, oriented parallel to the axes.
Rectangle[{xmin, ymin}, {xmax, ymax}, graphics] gives a
rectangle filled with the specified graphics.
```

Because the specification lists describe bounds for all of the data bins, they contain one more number than the data list. We can use **Drop** to remove the first or last item from the specification list, depending on which is appropriate for the minimum or maximum x value. To make the rectangles distinct, we can also add or subtract a small percentage of the maximum specification from the minimum and maximum values. The y values consist of zeros for the minimum and of the values of the original data for the maximum. The hardest part of the writing the plot function is determining how to rearrange our lists of values into a form that we can apply to **Rectangle**. The **Transpose** function works nicely to change a matrix of lists into a list of matrices – once we know which levels to swap. Here, we swap level 3 with level 1 and leave level 2 alone. Our graphics consist of the red color specification, **RGBColor[1,0,0]**, and a list of rectangles obtained by using **Apply** on the list boxes.

In:

```
BoxPlot[data_,spec_] :=
  Module[ {xmin,ymin,xmax,ymax,boxes},
          xmin = Drop[spec,-1] + 0.003 Max[spec];
          xmax = Drop[spec,1] - 0.003 Max[spec];
          ymin = Table[0.0, {Length[xmin]}];
          ymax = data;
          boxes = Transpose[ N[{{xmin,xmax},
                                {ymin,ymax}}],
                             {3,2,1}];
          Show[ Graphics[ {RGBColor[1,0,0],
          Apply[Rectangle, boxes,1]}],
          Axes->Automatic,
          AxesLabel->{"specs","qty"},
          PlotRange->All]
          ]

BoxPlot::usage="BoxPlot[data,binEdges] creates a
two-dimensional plot with rectangles representing the
data quantities in each bin. The rectangles are scaled
according to the bin widths. No options are supported."
```

Now **BoxPlot** shows rectangle heights consistent with our data and with widths given by the specification list.

In:

    BoxPlot[bc, spec];

Out:

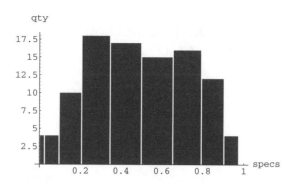

We can extend our work on histograms to make a three-dimensional box plot for viewing data with two parameters. Examining two parameters works well for discovering yield trade-offs in semiconductor devices, such as the resistance and capacitance of semiconductor diodes, or gain and noise figure of transistors. The help message for **BinCounts** shows that we can use it for two-dimensional histograms as long as the bin widths are uniform.

In:

    ?BinCounts

Out:

    BinCounts[data, {xmin, xmax, dx}] gives a list of the
    number of elements in data that lie in bins from xmin
    to xmax in steps of dx. BinCounts[{{x1,y1},
    {x2,y2}, ...}, {xmin, xmax, dx}, {ymin, ymax, dy}]
    gives an array of bin counts.

The following data contain noise figure and gain information for a few transistors.

In:

    dng = {{1.52,7.97},{1.6,7.95},{1.63,7.96},{1.65,8.03},
    {1.67,8.27},{1.56,8.24},{1.65,7.92},{1.66,7.85},
    {1.57,7.88}};
    b2 = BinCounts[dng, {1.25,1.75,0.05}, {7.5,8.5,0.1}]

Out:

    {{0, 0, 0, 0, 0, 0, 0, 0, 0, 0}, {0, 0, 0, 0, 0, 0, 0,
    0, 0, 0}, {0, 0, 0, 0, 0, 0, 0, 0, 0, 0}, {0, 0, 0, 0,
    0, 0, 0, 0, 0, 0}, {0, 0, 0, 0, 0, 0, 0, 0, 0, 0}, {0,
    0, 0, 0, 1, 0, 0, 0, 0, 0}, {0, 0, 0, 1, 0, 0, 0, 1,
    0, 0}, {0, 0, 0, 0, 3, 1, 0, 0, 0, 0}, {0, 0, 0, 1, 0,
    0, 0, 1, 0, 0}, {0, 0, 0, 0, 0, 0, 0, 0, 0, 0}}

BarChart3D will display this two-dimensional array as a typical Manhattan plot.

In:

```
Needs["Graphics`Graphics3D`"];
BarChart3D[ b2 ];
```

Out:

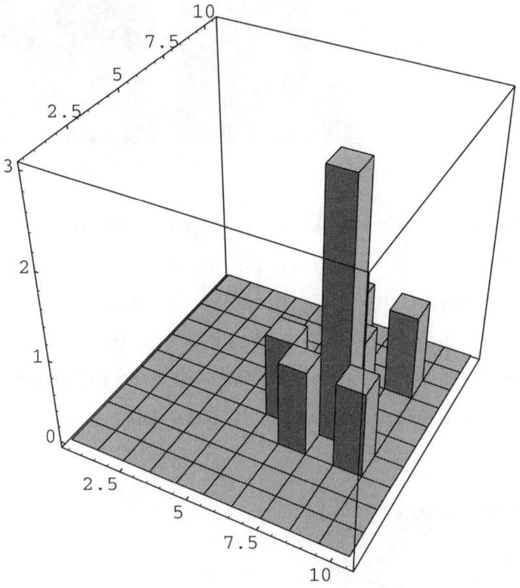

If we want nonuniform grids, we need more generalized functions. Our **histo2** function operates on data consisting of pairs, and two specification lists are used to bin the data into a matrix. **histo2** works in a similar manner to **histo1** by using **Select** to trim the data, **Length** to count the bins, and **Table** to manage what goes where.

In:

```
histo2[data_List,spec_List,spec2_List] :=
  Module[{i,j},
    Table[Length[ Select[Select[
      Select[Select[data,(#[[1]]<=spec[[i+1]])&],
             (#[[1]] > spec[[i]])& ],
             (#[[2]]<=spec2[[j+1]])&],
             (#[[2]] > spec2[[j]])& ]],
        {i,Length[spec]-1},
        {j,Length[spec2]-1}]
  ]
```

The following specifications set the noise figure and gain bins.

In:

```
nfs = {1.25, 1.35,1.45,1.55,1.6,1.65,1.75};
gs = {7.5,7.8,7.9,8.0,8.1,8.2,8.5};
dm = histo2[dng,nfs,gs]
```

Out:

```
{{0, 0, 0, 0, 0, 0}, {0, 0, 0, 0, 0, 0}, {0, 0, 1, 0,
0, 0}, {0, 1, 1, 0, 0, 1}, {0, 0, 2, 1, 0, 0}, {0, 1,
0, 0, 0, 1}}
```

Again, we can show these results with **BarChart3D**, but the bin areas are misrepresented.

In:

```
BarChart3D[ dm ];
```

Out:

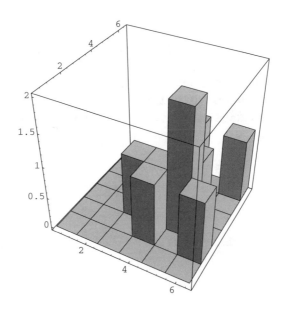

What we would really like to do is create three-dimensional boxes for each of the bins. Because we need the *z*-axis information, we will use **Polygon** for the graphics primitive. **Polygon** is more complicated to use than **Rectangle** because we need to specify all of the corner points instead of just the two opposite corners. We can take the two limit-specifications for the rectangle corners, {*xmin*, *xmax*} and {*ymin*, *ymax*}, along with the polygon height, *z*, and create a series of points using the following function. The **AntiOuter** function is similar to **Outer**, except that the ordering of the points is different – hence our choice of name.

In:

```
AntiOuter[{x1_,x2_},{y1_,y2_},z_] :=
  {{x1,y1,z},{x2,y1,z},{x2,y2,z},{x1,y2,z}}
```

In the following **box3D** function we use **Table** to pass all of the specifications through the **AntiOuter** function. Once we have all of the polygon points, we can map **Polygon** over the points and use **Graphics3D** to show the points. Note that because **Table** created a matrix of points, we had to use **Flatten[..., 1]** to create a long list of polygons for the mapping.

In:

```
box3D[data_,spec1_,spec2_] :=
  Module[ {tmp,boxes,polys,boxPolys,i,auxData},
  (* convert specs & data to {x,y,z} polygons *)
  polys = Flatten[Table[ AntiOuter[
            spec1[[{i,i+1}]],spec2[[{j,j+1}]],
            data[[i,j]] ],
            {i,Length[spec1]-1},
            {j,Length[spec2]-1}],
        1];
  Show[ Graphics3D[ {
  Polygon /@ polys}],
  BoxRatios->{1,1,1},PlotRange->All]
  ]
```

If we use **box3D**, as shown below, we get only the tops of our boxes – but already we can see the effect we are trying to create.

In:

```
    box3D[dm,nfs,gs];
```
Out:

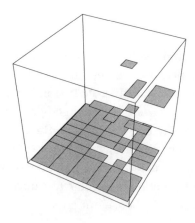

The complete Manhattan plot needs walls on the side of the boxes. To finish off the boxes, we first collect those polygons with heights above zero by selecting those polygons with their z point, **#[[3]]**, greater than zero. (You will notice we use **Flatten[...,1]** regularly to put lists at the right level for other manipulations.) The second instance of boxes manipulates the points from a long list of points, through a grouping of four points, to a grouping of two points with an offset of one. Then we use the **Append** function to create a fifth point, which closes the polygon. Finally, **Join** and **Reverse** complete the points list and set it up for one last partitioning of points into groups of four. Now **Graphics3D** also includes a mapping for the box sides and sets up the axes and axes labels.

In:

```
BoxPlot3D[data_,spec1_,spec2_] :=
  Module[ {tmp,boxes,polys,boxPolys,i,auxData},
  (*    transpose data if specs in wrong order *)
  auxData = If[ Length[data] == Length[spec1]-1,
  data,Transpose[data] ];
  (*    convert specs & data to {x,y,z} polygons *)
  polys = Flatten[Table[ AntiOuter[
            spec1[[{i,i+1}]],spec2[[{j,j+1}]]],
            auxData[[i,j]] ],
            {i,Length[spec1]-1},
            {j,Length[spec2]-1}],
          1];
  boxes = Select[Flatten[polys,1],
            (#[[3]]>0)&]; (* get those > 0 *)
    (*
    partition the list into 4 point groups
    and after adding the fifth point (the
    polygon starting point) we create
    sublists of two points and an offset
    of one point
    *)
  boxes = Flatten[Partition[Append[ #,#[[1]] ],2,1]& /@
                  Partition[boxes,4], 1];
  boxPolys = Partition[Flatten[Transpose[ {
                    Join[ #[[1]],Reverse[#[[1]]] ],
                    Join[ #[[2]],Reverse[#[[2]]] ],
                    Join[ #[[3]],{0,0} ] }]& /@
                    (Transpose /@ boxes), 1],
                  4];
  Show[ Graphics3D[ {RGBColor[1,0,0],
        Polygon /@ polys,
```

```
                Polygon /@ boxPolys}],
                Axes->Automatic,
                AxesLabel->{"spec1","spec2","qty"},
                BoxRatios->{1,1,1},PlotRange->All]
        ]
```

In:

```
    td = BoxPlot3D[dm,nfs,gs];
```

Out:

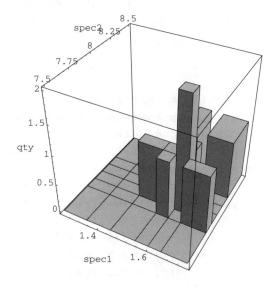

Now we have a full Manhattan plot, with all of the boxes scaled according to the bin widths for the specifications.

2.7   Acquisition-system problems

Data acquisition systems occasionally cause problems for the experimenter. Incoming signals can be corrupted by noise, and the acquisition system itself can malfunction. Exactly how you cope with these problems will depend on so many different factors that we cannot fully discuss the subject here. Instead, we hope that the following examples of problems and their solutions will provide a guide to appropriate manipulations.

2.7.1   Glitch removal

Our first problem example is a dataset that contains spurious negative values, typical of what might be returned by a data acquisition system that suffered an overload of some sort. If you were expecting data with values between 50 and 250, you might create a simple plot that would be messed up by the glitches:

In:

```
data={143, 194, 243, 162, 166, 111, -1, 158, 206, 127,
        102, 107, 182, 154, 138, 146, 138,
        127, 226, -1, 111, 226};
ListPlot[data,
          PlotJoined->True,
          PlotRange->{50,250}];
```

Out:

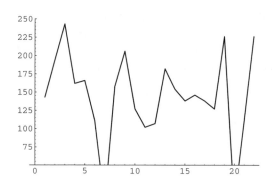

For such a small dataset, we could just edit out those erroneous values. If the dataset had several thousand points, then such an edit could be tedious. *Mathematica* can filter the data: we can map a function over these values that returns positive values unaltered and omits others. First, we declare an empty list **filtered** into which we shall collate the filtered data. Then we use an anonymous function (terminated like all such functions by an **&**) that tests to see if the current value (represented by the **#** symbol) it is processing is positive. If the test returns true then the current value is selected into the list **filtered**.

In:

```
filtered={};
filtered=Select[data, (#>0)&]
```

Out:

```
{143, 194, 243, 162, 166, 111, 158, 206, 127, 102, 107,
182, 154, 138, 146, 138, 127, 226, 111, 226}
```

By altering the test used in the **If** function, we can manipulate the data to take account of any other value-specific problem.

2.7.2 Cyclic errors

Many experimental analysis techniques assume that each member of a dataset has been acquired and manipulated identically. Sometimes, however, members are treated differently, and we might want to remove or compensate for those values in some way.

For example, a homemade data acquisition system takes ten samples each second. Because of a software feature (the processor spending time updating a clock card, say), the sample taken immediately after a clock second's "tick" is reduced in length by approximately 50%. Here are a few values from the corrupted dataset and a plot:

In:

```
data={1,0,0.5,2,3,4,5,6,7,8,9,10,5,
    12,13,14,15,16,17,18,19,20,9.5,
    18,17,16,15,14,13,12,11,10,4.5,8,7,6,5,4,3,2};
ListPlot[data,PlotJoined->True];
```

Out:

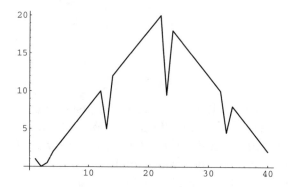

How we handle the corrupt dataset members will depend on many factors, including the extent to which we know the form of the corruption. The purest technique is to omit the corrupt values. Because the error is cyclic, we can use **Mod**, the modulus function, as part of a test to identify corrupt members by their position in the dataset. We can map a simple chooser function over our dataset. If a corrupt member is present, we substitute its numeric value by **Null**; uncorrupted members are left unaltered. When *Mathematica* encounters **Null** in a dataset, that point is not plotted.

In:

```
filtered={};
i=1;
Map[If[Mod[7+i++,10]!=0,
    AppendTo[filtered,#],
```

```
        AppendTo[filtered,Null]]&,
    data];
ListPlot[filtered,PlotJoined->True]
```

Out:

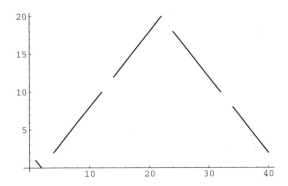

This technique will generate some warning messages of the form

Message out:

```
Graphics::gptn:
    Coordinate Null in {3, Null}
        is not a floating-point number.
```

but renders a truthful plot: no assumptions are made and the breaks in the dataset are clearly visible. If from this plot we think we now know how the data have been corrupted, we can attempt remedial action.

In:

```
filtered={};
i=1;
Map[If[Mod[7+i++,10]!=0,
        AppendTo[filtered,#],
        AppendTo[filtered,2 #]]&,
    data];
ListPlot[filtered,PlotJoined->True];
```

Out:

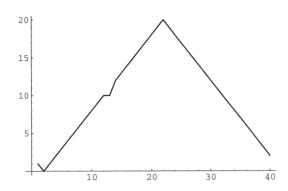

But sometimes we do not know as much as we think!

If we want to delete the corrupt members, we could omit the false clause in the **If** function, creating a shortened list that contains only good data (a technique applicable only if the data have no sequence- or order-sensitive attributes). If we want to replace corrupt members by their interpolated value, we need to fit a function to the dataset – a topic we discuss in Chapter 3.

2.7.3   Stuck ADC or encoder bits

You can see unexpected problems in data by plotting them in nonstandard ways. For example, computer systems used for acquiring analog data employ an analog-to-digital converter (ADC) to digitize the analog signal voltages. ADCs sometimes suffer from a problem that can be difficult to spot: one of the low-valued bits can get stuck in an *on* or an *off* state. A similar problem occurs with position encoders. With a few lines of *Mathematica* code, you can easily check for such a problem.

Here is a set of data that might have come from an experiment. By eye, it is very difficult to see that there is anything unusual in the numbers. Even plotting the numbers will not reveal any characteristic other than a general randomness.

In:

```
data={43, 94, 243, 62, 66, 111, 58, 206, 127,
      2, 7, 182, 54, 138, 146, 38,
      127, 226, 111, 226};
```

We can write a simple routine to convert a denary number to a binary number that we format as a list of bits:

In:

```
Clear[bits];
bits[x_]:=Module[{i,b={},c},
            c=x;
            For[i=7, i>=0,  i--,
                If[c>=(2^i),
                    c-=2^i;AppendTo[b,1],
                    AppendTo[b,0]];
                ];
            b]
```

Before we use the function, let us try it out:

In:

```
bits[4]
```

Out:

```
{0, 0, 0, 0, 0, 1, 0, 0}
```

We can then map the function bits over each member of the dataset and draw a three-dimensional bar chart to show, number by number, the state of each bit.

In:

```
bdata=Map[bits, data];
Needs["Graphics`Graphics3D`"];
BarChart3D[bdata,
        ViewPoint->{2.476, -1.399, 1.833},
        BoxRatios->{1,1,0.3},
        AxesLabel->{"sample","bit",""}
        ];
```

Out:

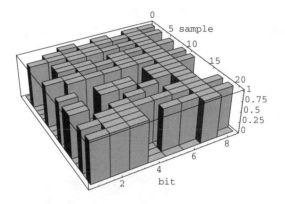

The bar chart shows clearly that, for all samples, one of the bits is on.

2.8   Listening to your data

Mathematica has the ability to create sounds from numbers, enabling you to hear your data. Sometimes, sets of data that look alike can sound quite different, and so the ability to create sound is definitely of practical use.

In addition to the sound-related functions in *Mathematica*, the package **Miscellaneous`Audio`** contains functions to play sounds that have simple, well-known waveforms (such as sine- or square-waves), to play amplitude- and frequency-modulated waveforms, and to specify various play parameters, such as the sampling rate and digitization depth. The package **Miscellaneous`Music`** contains functions that are scale- and tuning system-oriented.

2.8.1 Simple waveforms

As an introduction to sound, the functions **Waveform** and **Show** provide a quick way to try out your computer's sound system. Before we can use those functions, we need to load the audio package; we can then, say, generate a squarewave with a frequency of 1000 Hz that lasts for 2 seconds.

In:

```
Needs["Miscellaneous`Audio`"];
myWaveform=Waveform[Sawtooth,1000,2]
```

Out:

```
-Sound-
```

The function **Waveform** computes the numerical waveform and returns a Sound object that we can play by passing it to **Show**; the function **Show** is overloaded so that it works with both graphics and sound objects. The notion of showing a sound may seem strange, but *Mathematica* does also show a graphical representation of the sound:

In:

```
Show[myWaveform]
```

Out:

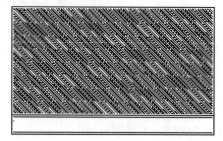

```
-Sound-
```

You can make your own waveforms and have them played by *Mathematica*. For example, you can generate a table of random numbers and hear them using **ListPlay**.

In:

```
myNoise=Table[Random[],{i,1,1000}];
ListPlay[myNoise]
```

Out:

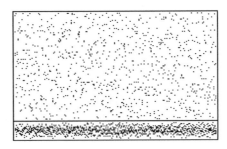

–Sound–

Just as we have used **ListPlay** to hear a table of random numbers, so you can use the same function to hear any list of experimental data, perhaps read in using **ReadList**.

You use **Play** to let you hear analytic functions (**Play** and **ListPlay** work analogously to **Plot** and **ListPlot**). Here is a 1-second 500-Hz sine wave:

In:

```
Play[ Sin[ 2Pi 500 t],{t,0,1}]
```

Out:

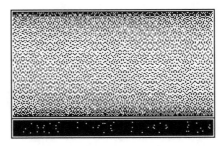

–Sound–

The duration of the sound produced by **Play** and **ListPlay** is proportional to the product of the number of items in the list and the time interval between successive samples. With lower sampling rates, the sound lasts longer but its frequency is lower.

In:

```
ListPlay[myNoise,SampleRate->1000]
```

Out:

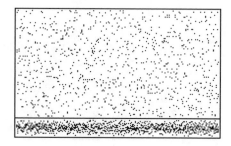

-Sound-

If you make the sampling rate too low, then Nyquist's sampling theorem is no longer satisfied. For example, sampling at a rate equal to the signal's frequency causes an identical value to be calculated for each sample. Sampling at a rate only a bit higher gives values each from a slightly earlier part of the sinewave than its predecessor. A 500-Hz sinewave sampled at 501 Hz results in a 1-Hz signal which is visible in the output graph but inaudible:

In:

```
Play[ Sin[ 2Pi 500 t],{t,0,1},SampleRate->501]
```

Out:

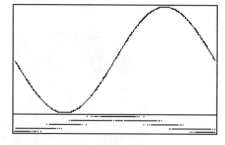

-Sound-

(If you connect an oscilloscope to the sound output of your computer, you can verify that a 1-Hz sinewave is indeed produced.)

2.8.2 Listening to the inaudible: Using modulation

Short lists or functions that have only low-frequency information are difficult to hear using **Play** or **ListPlay** directly. For example, a list of about 100 numbers sampled at 8000 Hz will only last just over 12 ms – barely recognizable as a bleep regardless of its frequency content.

In:

```
myList=Table[Abs[0.5-i],{i,0,1,0.01}];
ListPlay[myList]
```

Out:

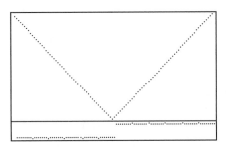

-Sound-

Modulation of a higher-frequency carrier signal by the low-frequency or short-length dataset allows you to hear the otherwise inaudible sounds. Our first example of modulation uses amplitude modulation: the amplitude of the carrier is multiplied by the lower-frequency dataset.

You can create the carrier signal using the **Table** function and check what the carrier sounds like by playing it with **ListPlay**. (You should try different sample rates, too.)

In:

```
myCarrier=Table[Sin[ 234 t],{t,0,2000}]//N;
ListPlay[myCarrier,SampleRate->1000]
```

Out:

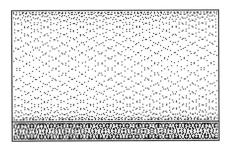

-Sound-

(Note: A 20,000-element carrier-signal list takes 3m 15s to generate on a Power Macintosh running at 60 MHz.)

Assuming that we have more carrier signal than modulating signal, using **Partition** we can break the long carrier list into sections, one for each value of the modulating signal, and multiply each carrier section by its associated modulating value. Because we want to ensure that the number of carrier sections is not larger than the number of modulating values, we use **Ceiling** to make each carrier section as long as possible.

In:

```
enoughCarrier=Partition[myCarrier,
                        Ceiling[Length[myCarrier]/
Length[myList]]];
```

Once the carrier signal has been broken down into its sections, we take the same number of modulating values as we have sections from our original dataset and multiply each carrier section by its modulating value. This gives us "modulated" sections that we have to destructure, or flatten, into one long list before we can invoke **ListPlay**.

In:

```
amSignal=Flatten[Take[myList,
                      Length[enoughCarrier]]
enoughCarrier];
ListPlay[amSignal,SampleRate->1000]
```

Out:

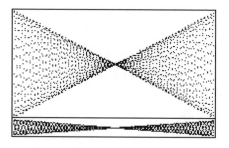

-Sound-

The graphical representation of **amSignal** shows how the carrier has been modulated in amplitude by the values in **myList**.

Frequency modulation is slightly harder to achieve because we need to know the values contained in the modulating signal when we generate the carrier. (The same requirement arises in radio transmission systems.)

In:

```
fmSignal=Table[ Sin[t 15/(1+myList[[f]])],
{f,1,Length[myList]},{t,0,2 Pi,2 Pi/100}]//Flatten;

ListPlay[fmSignal,SampleRate->2000]
```

Out:

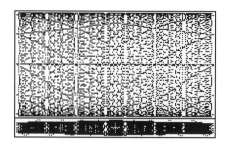

```
-Sound-
```

Differences in noise types are well indicated by the sound they make. For example, here are two sets of noise, the second of which has a longer correlation time and sounds less harsh. After you have listened to **noise1** and **noise2**, try plotting them using **ListPlot** and consider whether the difference is more striking when seen or when heard.

In:

```
noise1=Table[Random[],{i,1,1000}];
noise2=(noise1+RotateRight[noise1])/2;
```

If we now play the two noises, one immediately after the other, we can compare them directly:

In:

```
Flatten[{noise1,noise2}];
ListPlay[%,SampleRate->1000]
```

Out:

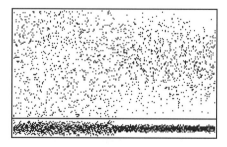

```
-Sound-
```

The medium of sound does not suit all types of data, so you need to determine whether or not it is appropriate to your application, just as you need to choose which type of graph to use when visualizing data.

2.9   References

Shaw, W. T., Tigg, J., "Applied *Mathematica*," Addison-Wesley, Reading, Massachusetts, USA, 1994.

The *Mathematica* Journal, Miller Freeman Inc., San Francisco, California, USA.

Wickham-Jones, T., "*Mathematica* Graphics," Springer-Verlag TELOS, Santa Clara, California, USA, 1994.

Wolfram, S., "*Mathematica* — A System for Doing Mathematics by Computer," Addison-Wesley, Redwood City, California, USA, 1991.

CHAPTER 3

Data analysis

Data analysis is at the heart of the scientific process: observe, analyze, hypothesize, experiment. Engineers, too, need to analyze data to work out why, say, some machine has misperformed. Whatever the professional discipline, if you do not analyze the available data, you are not able to proceed logically to the next hypothesis or experiment design.

Analysis, of course, has two phases: first, you have to design the test you want to apply to the data, and second, you have to execute that test. For example, if you are testing the relationship between how high a plant grows each day and how much nutrient it is fed, you need to decide how complicated that function will be; growth rate might be proportional to the volume of nutrient dispensed, or to the square-root of the volume – but it is unlikely to be proportional to the sine or cosine of the volume.

Mathematica can help you to display the characteristics of the data or to work out whether your choice of function is good, but, of course, you alone can provide the inspiration for the design of your tests and the functions that you fit to your data. In this chapter, we look at using *Mathematica* to carry out statistical analysis of data, to graphically compare theory and data, and to fit theoretical functions to data.

3.1 Statistical characteristics

The standard package **Statistics `DescriptiveStatistics`** contains many useful functions for calculating common statistical characteristics of data. We summarize these functions in Table 3:i but we go no further to avoid simply repeating the details that appear in the package documentation (both paper and online versions) that is supplied with *Mathematica*.

In addition to the functions listed in Table 3:i there are the functions **LocationReport[***data***]**, **ShapeReport[***data***]** and **DispersionReport[***data***]**, which return summaries of their arguments.

Table 3:i Functions in **Statistics`DescriptiveStatistics`** (alphabetically listed)

Function	Related function(s)
CentralMoment[*data,r*]	
InterquartileRange[*data*]	
Kurtosis[*data*]	KurtosisExcess[*data*]
Mean[*data*]	GeometricMean[*data*]
	HarmonicMean[*data*]
	TrimmedMean[*data,f*] and
	TrimmedMean[*data, {f1, f2}*]
Median[*data*]	
Mode[*data*]	
Quantile[*data, q*]	InterpolatedQuantile[*data, q*]
Quartiles[*data*]	QuartileDeviation[*data*]
	QuartileSkewness[*data*]
RootMeanSquare[*data*]	
SampleRange[*data*]	
Skewness[*data*]	PearsonSkewness1[*data*]
	PearsonSkewness2[*data*]
StandardDeviation[*data*]	StandardDeviationMLE[*data*]
StandardErrorOfSampleMean[*data*]	
Variance[*data*]	VarianceMLE[*data*]
	VarianceOfSampleMean[*data*]

In:

```
Needs["Statistics`DescriptiveStatistics`"];
areas={591., 266.8, 158.7, 147., 121.6,
       114., 110.6, 104.1, 97.8, 97.1};
LocationReport[areas]
```

Out:

```
{Mean -> 180.87, HarmonicMean -> 133.849,
  Median -> 117.8}
```

One of the advantages of using a computer-algebra system is that you can see exactly how any statistical function works simply by applying it to a list of symbolic variables. For example, how does *Mathematica* calculate the harmonic mean and the standard error of the mean?

In:

```
d={a,b,c};
HarmonicMean[d]
```

Out:

```
      3
  ---------
  1   1   1
  - + - + -
  a   b   c
```

In:

StandardErrorOfSampleMean[d]

Out:

```
             -a - b - c 2        -a - b - c 2       -a - b - c 2
Sqrt[(a + ------------)  + (b + ------------)  + (------------ + c) ]
               3                     3                  3
-------------------------------------------------------------------
                           Sqrt[6]
```

You can also see how *Mathematica* builds up functions from those already defined by wrapping the argument you supply with **Hold** to prevent the function evaluating the argument.

In:

StandardDeviation[Hold[d]]

Out:

```
Sqrt[Variance[Hold[d]]]
```

3.1.1 Preparing data

Many of *Mathematica*'s statistical functions expect a list of numbers, **{1.2, 3.4, 1.3,...}**. If you have data that are contained within some kind of structure, then you will need first to extract the numbers on which you want a statistical function to act. For example, you may have a list of number pairs where the first number of each pair is some kind of marker and the second number is the value in which you are interested. **Mean** cannot act on such a list.

In:

dList={{a1,b1},{a2,b2},{a3,b3}};
Mean[dList]

Out:

```
Mean[{{a1, b1}, {a2, b2}, {a3, b3}}]
```

The simplest way to extract the second number of each pair is to map a pure function over the list and then take the mean of the result (a list of single numbers). The pure function takes the second part of each pair to which it is applied.

In:

```
Map[Part[#,2]&, dList]
```

Out:

```
{b1, b2, b3}
```

In:

```
Mean[%]
```

Out:

```
b1 + b2 + b3
------------
     3
```

You can also use pure functions to apply any preprocessing that you want to do. For example, if you wanted to calculate the mean of the squares of the first number in each pair of **dList**, you might proceed as follows.

In:

```
Mean[Map[Part[#,1]^2&, dList]]
```

Out:

```
  2     2     2
a1  + a2  + a3
--------------
      3
```

3.1.2   Statistics without numbers

Because *Mathematica* can work symbolically, you can create plots to show the statistical characteristics of nonnumeric data. Here is a list of finds from an archaeological excavation – but it also could have been a list of plants found on a botanical expedition, or of items sold during an aerospace trade exhibition.

In:

```
finds={"jug","pot","jug","cup","beaker"};
```

Some statistical operations, such as calculating the mean, are not meaningful on nonnumeric data, but other operations are. For example, you can determine the most frequently found object using **Mode**:

In:

```
Mode[finds]
```

Out:

```
jug
```

By loading two more packages, you can generate bar and pie charts that show the frequency with which each object was found.

The data manipulation package provides many functions that can extract data from arrays (that is, by choosing a specific column or row) and functions that can choose data that pass a given test. Here, we use the **Frequencies** function to count for us how many times the objects occur in **finds**.

In:

```
Needs["Statistics`DataManipulation`"];
howOften=Frequencies[finds]
```

Out:

```
{{1, beaker}, {1, cup}, {2, jug}, {1, pot}}
```

We can then feed the result **howOften** directly into the **BarChart** and **PieChart** functions that are contained in one of the graphics packages.

In:

```
Needs["Graphics`Graphics`"];
BarChart[howOften]
```

Out:

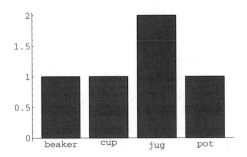

In:

```
PieChart[howOften];
```

Out:

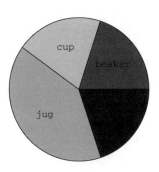

3.2   Comparing theory and data

Scientific analysis of empirical data often relies on comparing those data with values calculated from theory. In this instance, by theory we mean either a well-established theory or law, such as Ohm's law, or your freshly invented hypothesis.

3.2.1   Generating theoretical results

Mathematica provides both procedural and functional methods for generating values from theoretical functions. The procedural method is typical of the approach taken in traditional computer languages: a loop structure is executed as many times as required to produce data that span the same region of some parameter space as those data from the experiment. *Mathematica* has three main loop constructs: **For**, **Do**, and **While**. These functions work in basically the same way as they do in other computer languages. You need variables to be altered within the loop, their starting values, a test that returns false when you want the loop to terminate, and a specification for what happens to the loop variable during each execution of the loop.

In each case below, we declare an empty list, **{}**, and append to that list the number pair **x** and **1+x+x^2**, so giving a list of $\{x, y\}$ coordinates.

In:

```
myTheoryA={};
For[x=1,
    x<=6,
    x=x+1,
    AppendTo[myTheoryA,{x,1+x+x^2}]
  ];
myTheoryA
```

Out:

```
{{1, 3}, {2, 7}, {3, 13}, {4, 21}, {5, 31}, {6, 43}}
```

In:

```
myTheoryB={};
Do[AppendTo[myTheoryB,{x,1+x+x^2}],
   {x, 1, 6, 0.5}
  ];
myTheoryB
```

Out:

```
{{1, 3}, {1.5, 4.75}, {2., 7.}, {2.5, 9.75}, {3., 13.},
{3.5, 16.75},{4., 21.}, {4.5, 25.75}, {5., 31.}, {5.5,
36.75}, {6., 43.}}
```

In:

```
myTheoryC={};
x=1;
While[x<=6,
        AppendTo[myTheoryC,{x,1+x+x^2}];
        x=x+2
        ];
myTheoryC
```

Out:

```
{{1, 3}, {3, 13}, {5, 31}}
```

If you are transitioning to *Mathematica* from another computer language, there are several points to note.

- The *Mathematica* **For** loop requires a test (for example, **x<=6**) that if True allows execution of the loop, similar to C or C++; in Ada, BASIC, FOR-TRAN, and Pascal, only the terminating value of a **FOR** structure is required (for example, **FOR X=1 TO 6**). *Mathematica*'s **For** loop is similar to the same structure in C except that the role of commas and semicolons is reversed.

- The *Mathematica* **Do** loop is different to the FORTRAN **DO** structure in that the body of the loop is specified before the start, end, and increment of the controlling variable.

- In *Mathematica*, the body of the loop is contained within one argument of the loop function; in most other languages, it is not necessary to think about the body as being an argument of the loop structure because the loop end marker is a separate program statement such as **END**, **NEXT**, **END LOOP**, or **ENDDO**.

- Where the body of a loop, a start specification, a test, or the action to be taken on each loop repetition contains more than one line, *Mathematica* separates those lines by semicolons. In C, such multiple lines are separated by commas (and are not applicable in most other languages).

Functional programming in *Mathematica* allows you to calculate discrete values of a function or, for the purposes of plotting a function, to completely ignore the need to generate values that are subsequently plotted. You can think of **myTheoryB**, above, as a table of values – so *Mathematica*'s **Table** function is what you use:

In:

```
myTheoryD=Table[{x,1+x+x^2},
                {x,1,6,0.5}]
```

Out:

```
{{1, 3}, {1.5, 4.75}, {2., 7.}, {2.5, 9.75}, {3., 13.},
 {3.5, 16.75}, {4., 21.}, {4.5, 25.75}, {5., 31.}, {5.5,
 36.75}, {6., 43.}}
```

For plotting functions, you just use **Plot**. You do not need to precompute
values that you then plot as a separate exercise:

In:

```
myTheoryPlot=Plot[1+x+x^2,
                {x,-5,5},
                AxesLabel->{"x","f(x)"}];
```

Out:

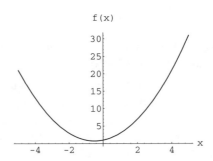

3.2.2   Plotting theory and data

Given that you have some empirical data and a theoretical function, a pre-
liminary task is to plot them together on the same plot, just so that you can
see how well they match. In our previous section, we discussed how you
can generate tables of values; in this section we look at displaying the theory
and the data together.

For example, from some experiment we have a dataset **myData**. Here is
a plot of **myData**:

In:

```
myDataPlot=ListPlot[myData,
                AxesLabel->{"x","observed f(x)"},
                PlotRange->All]
```

Out:

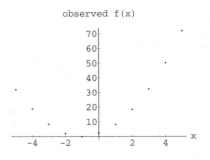

We can use **Show** to display these data and our theoretical function on the same plot, or we can use **GraphicsArray** to show them side by side.

In:

    Show[myDataPlot,myTheoryPlot];

Out:

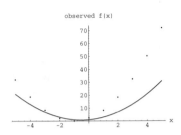

In:

    Show[GraphicsArray[{myDataPlot,myTheoryPlot}]];

Out:

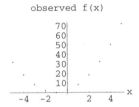 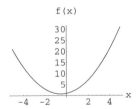

You might want to graphically compare your data with several theoretically derived plots. For example, you can generate a family of plots of the function $1 + x + a1\, x^2$, where the constant $a1$ ranges from 1 to 5 by giving **Plot** a list of functions to plot over the range $-5 \le x \le 5$. One way to generate the list is to use the replacement operator (**/.**) to make five versions of the function, with **a1** taking the values **1** through **5**. You must also make sure that *Mathematica* evaluates the function prior to plotting, or else it will try to plot a function with the (unevaluated) replacement rule still present.

In:

    Clear[a1,x];
    myTheoryPlot2=Plot[Evaluate[Table[(1+x+a1*(x^2)),
                                  {a1,1,5}]],
                 {x,-5,5},
                 AxesLabel->{"x","f(x)"}];

Out:

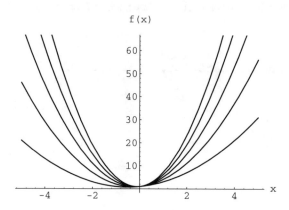

In:

```
Show[myDataPlot,myTheoryPlot2];
```

Out:

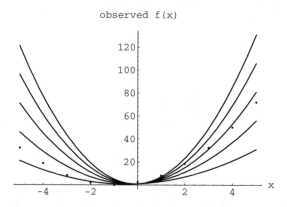

We can make this plot easier to interpret if we label each of the theoretical lines with the parameter used in its generation. The simplest way to make a label is to use the function **Text** **[label, {x, y}]** to place a label at a given **{x, y}** location. We choose to take the location for each label as that just after the positive-*x* end of the curve. (Note that we make *Mathematica* evaluate the curve's equation for us.) Once we have generated the labels, we can use **Show** to combine the data and theory plots with the labels, which we convert to graphical form using **Graphics**.

In:

```
labels=Table[Text[a1,
            ReplaceAll[{x,(1+x+a1(x^2))},
                    x->5.1]],
        {a1,1,5}]
```

Out:

```
{Text[1, {5.1, 32.11}], Text[2, {5.1, 58.12}], Text[3,
{5.1, 84.13}], Text[4, {5.1, 110.14}],Text[5, {5.1,
136.15}]}
```

In:

```
Show[myDataPlot,
    myTheoryPlot2,
    Graphics[labels]];
```

Out:

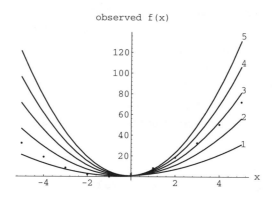

3.2.3 Plotting theory–data differences

The graphical comparison provides some indication of how well a function, or group of functions, might fit a dataset. For a more quantitative approach, you can calculate the numerical difference between the data and some chosen function at each data point. Your first task is to extract from the data all the *x*-values at which data were gathered and then to evaluate the theoretical function for each of those points, to form a difference, and to plot the differences.

The dataset **myData** is a list of *x*, and *f(x)* values, so you can extract the set of *x*- and *y*-values by extracting the first and second parts of each pair:

In:

```
xValue=Map[Part[#,1]&, myData]
```

Out:

```
{-5, -4, -3, -2, -1, 0, 1, 2, 3, 4, 5}
```

In:

```
yValue=Map[Part[#,2]&, myData]
```

Out:

```
{32.1, 18.8, 8.47, 2.05, -0.000187, 2.51, 8.28, 18.5,
32.5, 50.3, 72.1}
```

You can calculate theoretical function (**1+x+2x^2**) values for each of the *x* values read in, which you can then plot.

In:

```
fyValue=1+x+2x^2 /. x->xValue
```

Out:

```
{46, 29, 16, 7, 2, 1, 4, 11, 22, 37, 56}
```

If you use **ListPlot**, that function will assume that your data have an *x*-coordinate that runs from **1** to **11**, rather than over the correct range of **–5** to **+5**. One way of creating a list of correctly valued {*x, y*} pairs for plotting is to map a pure function over **fyValue** such that a pair of numbers is created for each number in **fyValue**. The first number of the pair is an *x*-value to which we give an initial value of −5 (and then increment that number after each use); the second number is the corresponding *y*-value.

In:

```
x=-5;
fxy=Map[{x++,#}&, fyValue]
```

Out:

```
{{-5, 46}, {-4, 29}, {-3, 16}, {-2, 7}, {-1, 2},
{0, 1}, {1, 4}, {2, 11}, {3, 22}, {4, 37}, {5, 56}}
```

In:

```
myTheoryPlot3=ListPlot[fxy,
                AxesLabel->{"x","y"},
                PlotJoined->True];
```

Out:

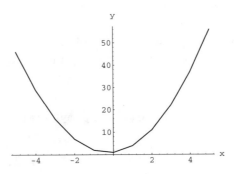

To form the difference at any *x*, you subtract the computed value from the observed value. The simplest way to plot the differences is to make an

$\{x, dy\}$ pair and then to use **ListPlot**. At present, we have two arrays: one contains the x-values and the other the observed–computed (o–c) differences. To use **ListPlot**, we need an array of $\{x, o$–$c\}$ pairs, which is the transpose of the array $\{$**xValue**, **differences**$\}$.

In:

```
differences=yValue-fyValue
```

Out:

```
{-13.9, -10.2, -7.53, -4.95, -2.00019, 1.51, 4.28, 7.5,
10.5, 13.3, 16.1}
```

In:

```
xyArray={xValue,differences}
```

Out:

```
{{-5, -4, -3, -2, -1, 0, 1, 2, 3, 4, 5}, {-13.9, -10.2,
-7.53, -4.95, -2.00019, 1.51, 4.28, 7.5, 10.5, 13.3,
16.1}}
```

In:

```
xyPairs=Transpose[xyArray]
```

Out:

```
{{-5, -13.9}, {-4, -10.2}, {-3, -7.53}, {-2, -4.95},
{-1, -2.00019}, {0, 1.51}, {1, 4.28}, {2, 7.5},
{3, 10.5}, {4,13.3},{5, 16.1}}
```

In:

```
residualsPlot=ListPlot[xyPairs,
        AxesLabel->{"x","o-c difference"}];
```

Out:

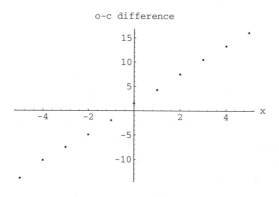

From this plot of the differences, or residuals, it appears that the residual is proportional to x, so perhaps an alteration to the coefficient of the x-term would be sufficient to give us a usable function with which to model our data.

Although this elementary analysis and residual plotting has not provided us with an immediate answer, it has shown that, for the range we are considering, a simple polynomial of degree 2 may suffice – there is certainly no indication that the function we seek is, say, a sine-like function.

3.3   Fitting functions to data

Mathematica contains a number of functions that will give best estimates of functions that fit a given dataset. Before we discuss them, you need to be aware of definitions and characteristics of different methods; function fitting is a subject with many traps for the unwary.

For example, the set of points {{1.6, 1}, {7.8, 1}, {14.1, 1}} can be fitted at least by the straight line $y = 1$, by the trigonometric function $sin(x)$, or (unjustifiably because there are only three points) by a family of polynomials of degree 3 or higher. As you can see, interpolating between fitted points or extrapolating beyond the range of variables given could give you unexpected results, so choosing the function to fit is an important task for you to do.

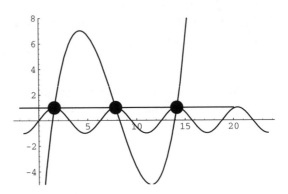

3.3.1   *Mathematica's* built-in fitting functions

There are two main functions in *Mathematica* that can help you fit functions to your data. One performs linear fitting, and the other performs nonlinear fitting. Understanding the difference between "linear" and "nonlinear" is important because it enables you to correctly choose which function to use.

A function f is linear in some coefficient c if $y = f(x)$ and $cy = f(cx)$. For example, the equation $y = ax^2$ is linear in a, but $y = sin(ax)$ is nonlinear in a because $sin(ax) \neq a\,sin(x)$.

3.3.1.1 Linear fitting

The function **Fit** takes three arguments: data, the function to fit, and the variables used in the fit. **Fit** returns an equation that you can invert to provide you with data values calculated from **Fit**'s model to compare with the empirical data. For example, **cfData** is a list of readings taken from a Celsius and Fahrenheit thermometer with the aim of discovering how the two temperature scales are related and how to convert from Celsius to Fahrenheit. Each data pair contains a Celsius value and a Fahrenheit value recorded at the same temperature. In this case, the second number (Fahrenheit) is a function of a constant and first number (Celsius); that is, the function we fit is the sum of some constant (multiplied by 1) and another constant multiplied by **c**, where **c** is the given variable.

In:

```
cfData={{30,86},{16,60},{26,79},{-10,13}};
fEquation=Fit[cfData, {1,c}, c]
```

Out:

```
31.1658 + 1.82801 c
```

fEquation tells us that a Fahrenheit value can be found from a Celsius value by multiplying the latter by 1.8 and adding 31. As we would expect, any fit to empirical data is not going to give us a precise answer; here we have an error of roughly 1°F in the zero-point of the Fahrenheit scale. Once we have determined the equation, we need to determine how well our function fits the data by looking at the residuals when we calculate the Celsius values from their Fahrenheit equivalents.

One method we can use is to write a function that, given a Fahrenheit value, returns its Celsius equivalent, as calculated using **fEquation**. In **getC**, we use **Solve** to calculate the Celsius temperature **c** that, when evaluated using **fEquation**, will equal the Fahrenheit temperature given as **getC**'s argument.

In:

```
getC[f_]:=Solve[f==fEquation, c]
```

For example, here is the Celsius equivalent of 86°F, as calculated using **fEquation**:

In:

```
getC[86]
```

Out:

```
{{c -> 29.9966}}
```

To extract the numerical value, we need to disassemble the one-element list of rules returned by **getC**. It is easiest to see the structure of **getC**'s output by expressing it in full format:

In:

```
FullForm[%]
```

Out:

```
List[List[Rule[c, 29.99661971830986]]]
```

The result is a one-element list with the element being another one-element list with a rule relating **c** with the given numerical value. So, the second part of the rule in the first element of the second list (which is the first element of the first list) is the number we want. We can express this disassembly in long form as

In:

```
Part[Part[Part[%%,1],1],2]
```

Out:

```
29.9966
```

or in short form

In:

```
getC[86][[1,1,2]]
```

Out:

```
29.9966
```

(Note that we could have appended the disassembly onto the end of the definition of **getC**, but we felt it made comprehension easier to take this one step at a time.)

To determine the Celsius values for all the Fahrenheit values in **cfData**, we can map **getC** over **cfData** and then calculate the root mean square (RMS) deviation taking the following steps. First we create a list of the squares of the differences between the calculated and measured Celsius values. We sum that list by applying the **+** operator to the list. Then we calculate the mean by dividing by the number of samples (that is, the length of **cfData**). Finally, we take the square root of the mean of the squares: the RMS value.

In:

```
cfTable=Map[{#[[2]],#[[1]],getC[#[[2]]][[1,1,2]]}&,
         cfData]
```

Out:

```
{{86, 30, 29.9966}, {60, 16, 15.7735},
{79, 26, 26.1673}, {13, -10, -9.93746}}
```

In:

```
Sqrt[Apply[Plus,
            Map[(#[[2]]-#[[3]])^2 &,
                cfTable]
            ]/Length[cfData]
        ]
```

Out:

```
0.144233
```

Writing your own code for such an operation is good practice in *Mathematica* coding, but if you do not want to write code, you can opt to use as many of the built-in functions as possible:

In:

```
Column[cfTable,2]-Column[cfTable,3]
```

Out:

```
{0.00338028, 0.226479, -0.167324, -0.0625352}
```

In:

```
RootMeanSquare[%]
```

Out:

```
0.144233
```

You can use **Fit** to give estimated fitting parameters to more complicated functions. For example, **myTable** contains values from an arbitrary function to which some noise, in the form of random real numbers within the range **-2** to **+2**, has been added. **Fit** copes well with such a function in cases where there is a sufficiency of data.

In:

```
myTable=Table[{x,N[Random[Real,{-2,2}]+
              2.3+0.5Sin[x]+0.1Exp[-x]+5x^4,
            2]},
        {x,-5,5,0.1}];
ListPlot[myTable,AxesLabel->{"x","f(x)"}];
```

Out:

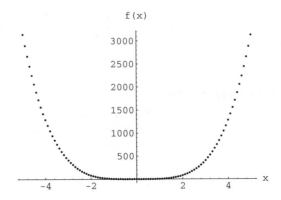

In:

```
Clear[x];
Fit[myTable,{1,Sin[x],Exp[x],x^4},{x}]
```

Out:

$$2.53123 - 0.0993496\ E^{x} + 5.02189\ x^{4} + 0.512057\ Sin[x]$$

If you try the same example with only a few (say, 5) data points, you can see that **Fit** still returns nearly correct values for the (linear) coefficients of the dominant terms in the function, but that the less significant coefficients can be quite poorly estimated. This is because the function has been undersampled for the amount of precision that we are trying to extract. Note that **Fit** does not warn you when its results have been degraded by undersampling. You will always find it a useful check to compare plots of your data with your **Fit**-derived function.

3.3.1.2 Nonlinear fitting

Many (simple) experiments result in data for which functions you may want to fit do not obey the criterion for coefficient linearity. The main set of *Mathematica* functions does not include one that can perform nonlinear fitting, but one such function is contained within the **Statistics** package set, which we now need to load.

In:

```
Needs["Statistics`NonlinearFit`"];
```

For example, the list **greenhouseTemp** contains two days' worth of temperature samples (taken hourly, in degrees Kelvin). Over the sampling period, what was the average temperature and the diurnal temperature variation?

In:

```
measuresPlot=ListPlot[greenhouseTemp,
                PlotJoined->True,
                AxesLabel->{"hours","t(K)"}];
```

Out:

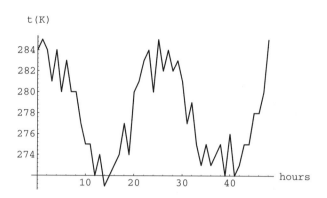

The function **NonlinearFit** takes four arguments: the raw data, the model for which we want to find a fit, the variable, and a list of parameters for which values are to be found. Here, we assume that the diurnal temperature variation can be modeled as a sine wave with some amplitude and phase offset. Although we know that the period of the diurnal variation is 24 hours, let us see if we can find that out from the data.

In:

```
model=tMean + tDiurnal Sin[f t + phase];
fitParameters=
    NonlinearFit[greenhouseTemp,
              model,
              t,
              {tMean,tDiurnal,f,phase},
              MaxIterations->100]
```

Out 2.2:

```
{tMean -> 278.027, tDiurnal -> 0.788449, f -> 0.817287,
phase -> 7.52659}
```

Out 3.0:

```
278.003 +1.17464 Sin[6.8644 +0.682971 t]
```

Let us see what this solution looks like. A quick look at the values given to **tDiurnal** implies that all may not be well; it is quite obvious that **tDiurnal** should be valued around 5. Also, we might expect the parameter **f** should be valued such that a complete sine-wave cycle will occur in 24 hours. That is, the **Sin** function will have an argument range from 0 to 2π as **t** ranges from 0 to 24. So

In:

```
2Pi/24//N
```

Out:

```
0.261799
```

should be close to **f** – not off by a factor of three, as it appears to be.

In 2.2:

```
fitPlot=
Plot[tMean+tDiurnal Sin[f t + phase]/.fitParameters,
     {t,0,48}];
```

In 3.0:

```
fitPlot=
Plot[fitParameters,          {t,0,48}];
```

Out:

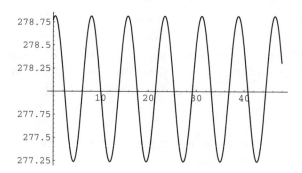

In:

```
Show[measuresPlot,fitPlot];
```

Out:

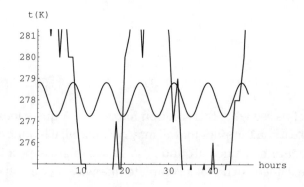

Clearly something is wrong. The fitting operation has probably failed in an effort to fit some of the noise in the data to a sine wave of higher frequency than that of the diurnal variation. Such a problem is not uncommon with numerical maximum- or minimum-hunting techniques. Press et al. (1990) and Michalewicz (1994) discuss the problems of hill-climbing algorithms in which a local extremum can mislead the algorithm into thinking it has found the correct solution (that is, the global extremum). You can get around such erroneous parochialism on the part of the algorithm either by giving it different starting conditions (sometimes called nudging) with the hope that the new starting values are closer to the correct solution, or by injecting some randomness into the process, as employed in the technique of simulated annealing (again see Press et al. and Michalewicz).

It is possible to nudge **NonlinearFit** by replacing the parameter to be nudged by a two-element list containing the parameter name and a suggested starting value. For example, here we try setting **f** to a value known to be quite close to the answer:

In:

```
fitParameters2=
    NonlinearFit[greenhouseTemp,
              model,
              t,
              {tMean,tDiurnal,{f,0.25},phase},
              MaxIterations->100]
```

Out 2.2:

```
{tMean -> 278.082, tDiurnal -> 5.63778, f -> 0.255339,
    phase -> 1.16867}
```

Out 3.0:

```
278.082 +5.63777 Sin[1.16864 +0.25534 t]
```

In 2.2:

```
fitPlot2=Plot[tMean+tDiurnal Sin[f t + phase]/.
    fitParameters2, {t,0,48}, DisplayFunction->Identity];
```

In 3.0:

```
fitPlot2=Plot[fitParameters2, {t,0,48},
    DisplayFunction->Identity];
```

Plotting the empirical data with our new fit function shows that we are much closer to the correct conclusion.

In:

```
Show[measuresPlot,fitPlot2];
```

Out:

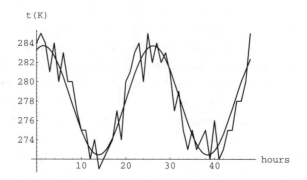

3.3.2 Fitting functions to complex data

Sometimes measurements represent complex numbers. Transfer function measurements can, for example, represent characteristics of electrical cables, electronic systems, and radar signal returns. Such measurements may be made in the time domain as impulse or step responses, or in the frequency domain as transfer functions. Often the result is more easily modeled as a frequency-domain transfer function, because a function with only a few poles or zeros can represent the entire waveform. How can we translate our complex numbers into a complex model?

Generally, we cannot obtain a satisfactory complex model from complex numbers by fitting one curve for the real part and a second curve for the imaginary part. Instead, most complex fits work in one of two ways. Either *i)* a series of moments of the impulse response is used to estimate the transfer function poles and zeros, or *ii)* the poles and zeros are fitted to the autocorrelation function of the signal. Fitting to the autocorrelation function corresponds to a moving average (MA) estimation of zeros, autoregressive (AR) estimation of poles, and ARMA estimation of both poles and zeros (see Tsui (1989)). Fitting poles and zeros to moments of the impulse response (which has been done for a long time) is called "asymptotic waveform evaluation" (AWE) (see Pillage et al. (1995)) and is the method upon which we will focus. Note that both of these methods average the data through integration and so reduce the effects of noise in the data.

The principal behind AWE methods is that the Laplace transform of the product of the impulse response **h[t]** and time **t** is related to the derivative of the transfer function **F[s]** through the following Laplace transform pair. (Here, multiplication by time is analogous to the first moment of the impulse response with respect to time.)

In:

```
L[t h[t]] = - D[F[s],s]
```

Also, the integration of the impulse response `h[t]` is related to the limit of the transfer function `F[s]` as `s` tends to zero:

In:

```
Integrate[h[t],{t,0,Infinity}] = Limit[F[s],s->0]
```

These two relationships give us a way to evaluate the transfer function by integrating a measurable response. We explore these techniques with two examples based on simple electronic circuits: a low-pass filter made from a resistor/capacitor (RC) circuit, and a band pass filter made from an inductor/capacitor (LC) resonator circuit.

3.3.2.1 Simple RC filter

Given an ideal filter with a known transfer-function pole, we can create a test impulse response by using the Fourier transform. The following function, **filterRC**, defines a simple RC-filter transfer function at frequency **f**. We also leave the time constant of the filter, **a**, as an argument.

In:

```
filterRC[f_,a_] := 1/(1 + (I 6.28 f a))
```

Next we create an array of positive and negative frequencies that we will turn into an impulse response by applying an inverse Fourier transform. The total length of the frequency array will be 256 points: 128 points of positive frequency, and 128 points of negative frequency. For this example, the frequency will range from 0 to 1280 Hz, in 10-Hz steps, and the full frequency-domain transfer function will be stored in the array **tf**.

In:

```
freqs = Range[0,1280,10];
sp = filterRC[freqs,0.0125];
sn = filterRC[-freqs,0.0125];
tf = Join[Drop[sp,-1], Reverse[Drop[sn,1]]];
```

We show the following transfer function plot in decibels to illustrate how the maximum attenuation factor is 0.01, or −40 dB. (We chose the **a** value to have this maximum attenuation to make an obvious filter.) The maximum attenuation appears at bin 128, and the folded-over negative frequencies appear in bins 129 through 256. If the frequency axis were on a logarithmic scale, the linear −20 dB/decade slope of this filter would be obvious.

In:

```
ListPlot[20.0 Log[10,Abs[tf]],PlotRange->All];
```
Out:

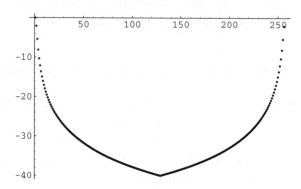

The **Fourier** function in *Mathematica* is not normalized, a feature that creates a problem when we analyze time-domain data to determine frequency-domain information. We can normalize the **Fourier** function by dividing by the square root of the number of points analyzed. So, the Fourier transform of **tf** is scaled by the square root of the number of points and stored in **impRC**. We take the real part of the result to remove any small-valued residual imaginary numbers, and we create the **tSteps** array so that we can have time values against which to plot the impulse response. Each time step is the inverse of the frequency step, 10 Hz, multiplied by the number of points in the transform (that is, the array length).

In:

```
impRC = Re[Fourier[tf]/Sqrt[Length[tf]]] //N;
tSteps=Range[0,Length[impRC]-1]/(10.0 Length[impRC]);
ListPlot[Transpose[{tSteps,impRC}],
        PlotRange->All,PlotLabel->"RC Impulse
        Response", AxesLabel->{"time","Vout"}];
```
Out:

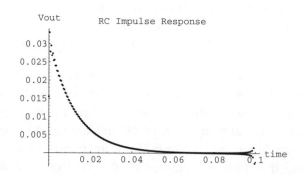

This does look like the impulse response for an RC filter. Note that we get some ringing in the impulse response around 0.1 second – the waveform begins to build back up – because the Fourier transform assumes the waveform is periodic. We should check that the integration of our impulse response gives us the limit of our transfer function as **s** approaches zero. This is a useful scaling check, especially when we know the answer should be unity.

In:

```
aCheck = Apply[Plus,impRC]
```
Out:

```
1.
```

One might expect to have to scale this summation, which is a simple approximation to an integration, by the time-step width. However, our step widths have already been factored into the height of the resulting waveform by the Fourier transform.

Lastly, we can estimate the first moment by multiplying the impulse response by our time-step array and summing the result. The answer is very close to our original **a** value of 0.0125, which we set in our filter response **filterRC**.

In:

```
ans = tSteps . impRC
```
Out:

```
0.012273
```

Later on we will compare answers derived from full numerical integration with answers from simple summations but, for now, a summation suffices.

We now apply the AWE method. How does this work? The Laplace transform of our original RC-filter transfer function is

In:

```
origRC = 1/(1 + s a);
```

The Laplace transform of the filter's first moment, **LaplaceTransform[t h(t), t, s]**, is:

In:

```
m1 = -D[ origRC, s]
```
Out:

```
        a
    ----------
            2
    (1 + a s)
```

We can effectively integrate this function, obtaining `Integrate [ t h(t),` `t]` by using the Laplace transform pair given at the beginning of this section. So, taking the limit of our transform, `m1`, as `s` approaches zero, effectively integrates our filter's first moment over all time.

In:

```
Limit[ m1, s->0]
```

Out:

```
a
```

So the first moment, which we computed by integrating (or summing) the product of the impulse response and the time array, gives us the time constant for the RC filter.

3.3.2.2 LC resonator

But what if the spectrum is more complicated? What if we wanted to find an approximation to a resonant circuit? A simple resonant circuit would be a shunt LC resonator driven by a source of resistance, `r`, and terminating in a resistor, `r`, and would have the following transfer function.

In:

```
tmp = z1/(r + z1) /. {z1->1/(1/r+sC+1/sL)} //Simplify
```

Out:

```
          sL
    ------------------
    r + 2 sL + r sC sL
```

If we divide through by `r`, we can get the transfer function with the time constants shown. Just so we can get a better feel for AWE, we should compare the resonator transfer-function extraction with a second-order low pass filter extraction. The following two transfer functions, `origRes` and `origLP2`, define second-order band pass and low pass filters, respectively.

In:

```
origRes = b1 s/(1 + a1 s + a2 s^2);
origLP2 = 1/(1 + a1 s + a2 s^2);
```

We can analytically derive the first moments of these functions by Laplace methods. The following analysis shows that the first moment of the band pass filter gives us the time constant of the zero, and that the first moment of the second-order low pass filter gives us a time constant similar to the one derived for the first-order filter. In all of the following cases, we can simply substitute zero for **s**, but you should use **Limit** if you suspect a singularity at zero.

In:

```
mr1 = -D[ origRes, s];
mLP1 = -D[ origLP2, s];
{mr1,mLP1} /. s->0
```

Out:

```
{-b1, a1}
```

We still have other time constants to compute, so we continue by creating higher-order moments.

The second moment of each transfer function contains two time constants. We use the first moment's result to extract our second time constant from the second moment.

In:

```
mr2 = -D[ mr1, s];
mLP2 = -D[ mLP1, s];
{mr2,mLP2} /. s->0
```

Out:

```
                 2
{-2 a1 b1, 2 a1   - 2 a2}
```

The band pass filter contains three time constants, so we should compute a third moment from its impulse response.

In:

```
mr3 = -D[ mr2, s];
mr3 /. s->0
```

Out:

```
        2
 -6 a1   b1 + 6 a2 b1
```

Each successive derivative gives us the information we need to extract details of the higher-order poles.

The main problem, of course, is that we need a reasonable model for our network. Here, we use a band pass filter to examine more complicated networks. The following transfer function **filterBP** describes a band-pass filter that has inductor and capacitor values that resonate at 200 Hz and have a Q of 1: $L = 39.8$ mH and $C = 15.8$ μF.

In:

```
filterBP[f_,c_,ind_,r_] :=
  Module[ {tp},
          tp = N[2 Pi I f];
          tp ind/r/(1+tp 2. ind/r + c ind tp^2)
        ]
bp = filterBP[freqs, 15.8 10^-6, 0.0398, 50.0];
bn = filterBP[-freqs, 15.8 10^-6, 0.0398, 50.0];
tfBP = Join[Drop[bp,-1], Reverse[Drop[bn,1]]];
```

In the plot of the band pass transfer function, **tfBP**, note both the folded over spectrum and the poor 11 dB of rejection at the highest analyzed frequency. We first plot the filter's spectrum in linear magnitude (not logarithmic dBs) to emphasize the band-pass characteristic and next the filter's impulse response:

In:

```
ListPlot[ Abs[ tfBP ], PlotRange->All,
          PlotLabel->"BandPass Spectrum"];
```

Out:

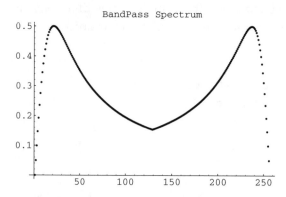

In:

```
impBP = Re[Fourier[tfBP]/Sqrt[Length[tfBP]]] //N;
ListPlot[Transpose[{tSteps,impBP}], PlotRange->All,
         PlotLabel->"BP Impulse Response",
         AxesLabel->{"time","Vout"}];
```

Out:

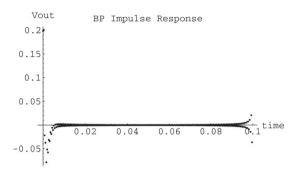

Note the ringing at the end of the impulse response, which is due to problems such as the limited high-frequency attenuation in our data. Because of both the small values in our data and the magnification of the ringing-induced errors by multiplying by large time values, it is better to truncate the dataset before it reaches these time values. We will truncate the dataset at the center because our oscillatory error is a minimum at the midpoint. We estimate **b1** by integrating the product of the impulse response and the time step array over the first half of the data.

In:

```
ans = Apply[Plus, Take[tSteps impBP,128]]
```

Out:

```
-0.000802363
```

Our actual value for **b1**, from our input inductance of 39.8 mH and resistance of 50 Ω, is as follows:

In:

```
0.0398/50.0
```

Out:

```
0.000796
```

The actual and derived values are very close – as long as we remember that the first moment was supposed to return the negative of **b1**. Because our first moment returns a negative value, we get a clue that our transfer function is not just a low-pass filter. All low-pass filters will return a positive first moment approximately equal to the delay of the filter (the Elmore delay).

We can find the **a1** value for this filter from the second moment, which we calculate in the same way:

In:

```
ans2 = Apply[Plus, Take[tSteps^2 impBP,128]]
```

Out:

$$-2.52733 \ 10^{-6}$$

This value should be **-2 a1 b1**, or, because **a1 = 2 b1**,

In:

```
-4 (0.0398/50.0)^2
```

Out:

$$-2.53446 \ 10^{-6}$$

Again, we have an excellent approximation to our time constant. Just how much would an improved integration scheme change our answer? We can use the **ListIntegrate** in the package **NumericalMath** to find out.

In:

```
Needs["NumericalMath`ListIntegrate`"]
ListIntegrate[ Take[tSteps impBP,128], 1]
```

Out:

```
-0.00077926
```

ListIntegrate does not improve this result because we have sources of errors in our data that are larger than the improvement given by any algorithmic efficacy in **ListIntegrate**.

If we extract **a2**, then we can compare our approximate transfer function to the original band-pass filter:

In:

```
ans3 = Apply[Plus, Take[tSteps^3 impBP,128]]
```

Out:

$$-8.79767 \ 10^{-9}$$

Now we can extract **b1**, **a1**, and **a2**. However, first we use the **Solve** function to derive **a2** from our third moment, **ans3**, using a formula that we derive from the value of **mr3** as **s** approaches zero.

In:

```
Solve[ Ans3 == -6 a1^2 b1 + 6 a2 b1, a2]
```

Out:

$$\left\{\left\{a2 \rightarrow \frac{Ans3 + 6 \ a1^2 \ b1}{6 \ b1}\right\}\right\}$$

In:

```
comp = {B1->-ans,
        A1-> ans2/(2 ans),
        A2->ans3/(-6 ans) + (ans2/(2 ans))^2}
```

Out:

```
{B1 -> 0.000802363, A1 -> 0.00157493,
                    -7
 A2 -> 6.52945 10    }
```

We can tabulate these values and compare them to the exact values.

In:

```
TableForm[
Transpose[{{0.0398/50.0,
           2 0.0398/50.0,
           0.0398 15.8 10^-6},
          {B1,A1,A2} /. comp}],
TableHeadings->{{B1,A1,A2},{"Real","Approx"}}]
```

Out:

```
     Real          Approx
B1   0.000796      0.000802363
A1   0.001592      0.00157493
              -7                -7
A2   6.2884 10     6.52945 10
```

The exact and approximate values look reasonably similar, but as the order of the filter gets larger, the pole locations become very sensitive to small changes in the time constants. To compare our approximate filter with our original band-pass filter we need a time-constant-based transfer function, which we call **filterAprx**. (Only the positive-frequency spectrum is shown in the following plots.)

In:

```
filterAprx[f_,b1_,a1_,a2_] :=
  Module[ {tp},
          tp = N[2 Pi I f];
          tp b1/(1+tp a1 + a2 tp^2)
        ]

{b1a,a1a,a2a} = {B1,A1,A2} /. comp;
Plot[{ Abs[filterBP[fr,
                    15.8 10^-6, 0.0398, 50.0]],
      Abs[filterAprx[fr,b1a,a1a,a2a]]},
  {fr,1,1000},
  AxesLabel->{"Hz","Abs"},
```

```
PlotLabel->"BP Filter vs. Moment Approx",
PlotStyle->{GrayLevel[0],
Dashing[{0.01,0.015}]}];
```

Out:

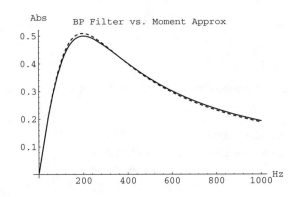

So our approximated transfer function does contain some errors, but they are small. What if we did not know how high an order our filter really was? If we approximate this band pass filter with just the first (or just the first and second) moments, we get the following responses.

In:

```
Plot[{ Abs[filterBP[fr, 15.8 10^-6, 0.0398, 50.0]],
       Abs[filterAprx[fr,b1a,0.0,0.0]],
       Abs[filterAprx[fr,b1a,a1a,0.0]]},
    {fr,1,1000},
    AxesLabel->{"Hz","Abs"},
    PlotLabel->"BP Filter vs. Lower Moments",
    PlotStyle->{GrayLevel[0],
    Dashing[{0.01,0.015}],
    Dashing[{0.05,0.02}]}];
```

Out:

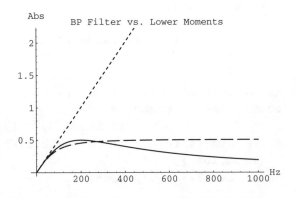

The solid line is our original band pass filter. The first-order approximation creates a high-pass response with a pole at infinity, which creates a large error above 200 Hz. The second-order approximation includes a zero at infinity, which cancels the pole and leaves a simple high-pass filter characteristic. For an accurate approximation above 400 Hz, we need the second zero at infinity.

3.3.3   Spline-based fitting

Data are not always fit to a curve whose analytic function is used directly. Sometimes we need the derivative of the function. For example, a varactor is a reverse-biased diode whose junction capacitance is a function of the bias voltage. Measurements of varactor capacitance-versus-voltage (CV) characteristics can give much more than simply what voltage you need to apply to achieve a certain capacitance. Derivatives give information on the device nonlinearities, and these can be used to construct Volterra-series nonlinear models for devices (Maas (1988)). As an example, we shall construct a CV curve and sample it numerically. Then we can fit to the curve using various methods and examine how well each method approximates the derivatives of the original curve. (The **g** parameter in the capacitance function relates to the doping profile in the varactor and is equal to **0.5** for an abrupt junction.)

In:

```
cIV[v_,g_] := 1/((1 - v)^g)
vs = Range[0,0.9,0.1];
cs = cIV[vs,0.5]
```

Out:

```
{1, 1.05409, 1.11803, 1.19523, 1.29099, 1.41421,
1.58114, 1.82574, 2.23607, 3.16228}
```

In:

```
data = Transpose[{vs,cs}];
```

Now we can fit the data with **Fit**, and **SplineFit**, and examine the derivatives. First, however, we need to load **SplineFit** from the **Numerical-Math** package.

In:

```
Needs["NumericalMath`SplineFit`"]
```

Splines are good for generating smooth curves from data points. A spline is analogous to the mechanical drawing spline which helps connect discrete points into a continuous curve, and so gives a type of interpolation. A simple connecting-the-dots technique looks poor to the eye and has mathematically discontinuous derivatives. Many splines are generated over a dataset. Books covering computational numerical techniques (for example, Press et al. (1994)) often cover only cubic splines, but *Mathematica's* **SplineFit** function also allows **CompositeBezier** and **Bezier** splines, as the following help message shows.

In:

> **?SplineFit**

Out:

> SplineFit[points, type] generates a SplineFunction
> object of the given type from the points. Supported
> types are Cubic, CompositeBezier, and Bezier.

First, we will fit the data with splines, as shown below for **Cubic** and **Bezier** cases.

In:

> **s = SplineFit[data, Cubic]**

Out:

> SplineFunction[Cubic,{0., 9.},<>]

In:

> **b = SplineFit[data, Bezier]**

Out:

> SplineFunction[Bezier,{0., 9.},<>]

The spline function returns two values, representing the x- and y-coordinates of the point. Notice that although it seems as if you should enter the x-coordinate and receive the y-value as an answer, the splines are scaled such that for the dataset with x from 0 to 0.9 you need to enter 10.0*0.9 to get the y-value for $x = 0.9$ returned, as the following example shows. (This scaling is hinted at by the ranges given in the returned spline functions **a** and **b**.) Also, because we want the y-value, we must remember to use the second value returned by the spline function.

In:

> **b[10 0.9]**

Out:

> {0.9, 3.16228}

Now we will compare the spline curves and derivatives to the curves generated by **Fit**.

In:

```
f3 = Fit[data, {1,x,x^2,x^3}, x]
```

Out:

$$0.95346 + 1.97669\ x - 5.87191\ x^2 + 7.00396\ x^3$$

In:

```
f5 = Fit[data, x^Range[0,5], x];
```

We need to plot all the curves to get an idea of the errors involved. The **Cubic** spline fits much better than the **Bezier** spline, in this situation. In the following plot the cubic spline uses an evenly dashed line; the Bezier curve uses an asymmetric dashing.

In:

```
Plot[{cIV[x,0.5], s[10 x][[2]], b[10 x][[2]]},
     {x,0,0.9},
     PlotLabel->"Fitted CV Curves - Splines",
     AxesLabel->{"volts","C"},
     PlotStyle->{GrayLevel[0],
     Dashing[{0.01,0.015}],
     Dashing[{0.05,0.02}]},
     Compiled->False];
```

Out:

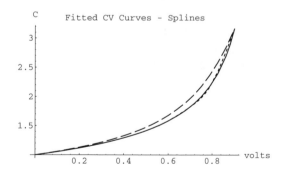

If we zoom in on the this plot, we can see how close the cubic spline is to the original data, and how the Bezier curve lies well above the original data.

In:

```
Show[%, PlotRange->{{0.6,0.8},{1.5,2.5}}];
```

Out:

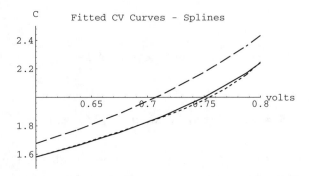

The polynomial curves computed by **Fit** show a gradual motion about the actual curve. The fifth-order curve shows a much smaller error and faster motion about the desired curve. If we extended the order of Fit to eight, we would get almost exact agreement to the original curve and the first and second derivatives. However, often polynomial fits will simply introduce more and more ripple, which allows all data points to be fit but creates significant errors in the derivatives, as shown at the beginning of Section 3.3. Generally, we would like to use a method which fits within the desired error, but with the lowest-order polynomial possible.

In:

```
Plot[{cIV[x,0.5], f3, f5}, {x,0,0.9},
PlotLabel->"Fitted CV Curves",
AxesLabel->{"volts","C"},
PlotStyle->{GrayLevel[0],
Dashing[{0.01,0.015}],
Dashing[{0.05,0.02}]}];
```

Out:

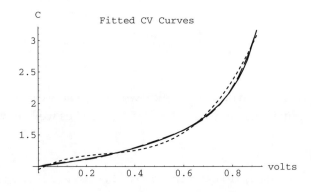

When we zoom in on the fitted curve, as shown in the following plot, we can see the fifth-order curve crossing the original curve more often than the third-order fit. (The third-order fit uses a finely dashed line, and the fifth-order curve uses longer dashes.)

In:

```
Show[%, PlotRange->{{0.6,0.8},{1.5,2.5}}];
```

Out:

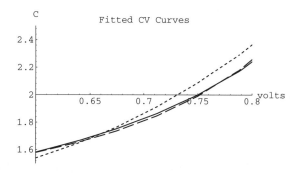

Our main concern in this example is how well the derivatives of the fitted curves agree with the derivatives of the original function. All of the polynomial fits and the original curve have analytic derivatives. As the following derivative of the third-order polynomial shows, we loose an order for each derivative.

In:

```
dc = D[cIV[x,0.5], x];
df3 = D[f3, x];
df5 = D[f5, x];
df3
```

Out:

$$1.97669 - 11.7438\ x + 21.0119\ x^2$$

Derivatives of *Mathematica*'s spline function are not so simple. We when try to take an analytic derivative of the spline function, we find that **D**, the *Mathematica* derivative function, cannot handle **SplineFunction**, as shown below. This is probably because each spline polynomial is valid over only a small range of the data. However, we can construct a numerical spline derivative, **splineD**, and use that function in our plots. Even though

splineD is approximate, it will still show how derivatives of splines differ from derivatives of polynomials that try to cover the entire dataset.

In:

```
splineD[fcn_, x_] :=
  Module[ {t1,t2},
          t1 = fcn[0.995 10 x][[2]];
          t2 = fcn[1.005 10 x][[2]];
          100 (t2-t1)/x
        ]
```

The spline first derivatives lie close to the first derivative of the original curve. Again, the cubic spline does a better job in this situation, although the Bezier curve gives a smoother function.

In:

```
Plot[{dc,splineD[s, x],splineD[b, x]},
     {x,0.05,0.85},
     PlotLabel->"Fitted CV 1stD - Spline",
     AxesLabel->{"volts","dC"},
     PlotStyle->{GrayLevel[0],
     Dashing[{0.01,0.015}],
     Dashing[{0.05,0.02}]}];
```

Out:

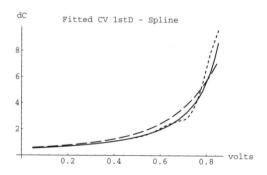

The polynomial first derivatives show larger errors and more obvious ripples than the spline derivatives. Because the polynomials from **Fit** try to cover the entire dataset, they are forced to oscillate about the desired curve in order to minimize the squared error.

In:

```
pf1 = Plot[{dc,df3,df5}, {x,0,0.9},
           PlotLabel->"Fitted CV 1stD",
           AxesLabel->{"volts","dC"},
           PlotStyle->{GrayLevel[0],
           Dashing[{0.01,0.015}],
           Dashing[{0.05,0.02}]}];
```

Out:

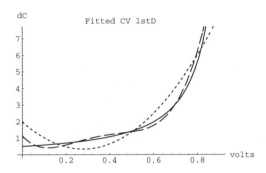

Second derivatives of the polynomial curves show even larger error. In fact, the third-order fit is a straight line! Although our polynomials are losing an order with each successive derivative, our original function maintains a fairly consistent shape when derivatives are taken.

In:

```
d2c = D[ dc, x];
d2f3 = D[df3, x];
d2f5 = D[df5, x];
d2f3
```

Out:

```
-11.7438 + 42.0238 x
```

In:

```
Plot[{d2c,d2f3,d2f5}, {x,0,0.9},
     PlotLabel->"Fitted CV 2ndD", AxesLabel->{"volts",
     "dC"}, PlotStyle->{GrayLevel[0], Dashing[{0.01,
     0.015}], Dashing[{0.05, 0.02}]}];
```

Out:

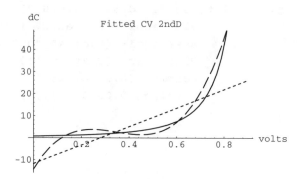

Generally one of two things happen when we use higher order fits: our fit gets better, or our fit begins to oscillate through the data points with disastrous results for the derivative information. At that point, we can use either a nonuniform data spacing to reduce the oscillation, or splines. A "Practical Guide to Splines" by de Boer (1978) details how choosing points according to a Chebyshev polynomial is nearly optimal and is preferred over equidistant points.

This next example shows how we can minimize the error in a region of a curve if we concentrate points in that region. We will put more points in the center of the curve and analyze the effect on the derivatives. The analysis is the same as the previous polynomial fit, except that most of the points are near the center of the x data.

In:

```
vns = {0.0,0.2,0.35,0.42, 0.44,0.46,
       0.48,0.55,0.7,0.9};
cns = cIV[vns,0.5];
datan = Transpose[{vns,cns}];
fn3 = Fit[datan, {1,x,x^2,x^3}, x];
fn5 = Fit[datan, x^Range[0,5], x];
```

The following plot of the nonuniform data fit shows results that are very similar to the uniform fit. Certainly we do not see large errors at the ends of the fit, even though there are fewer points at the ends now.

In:

```
Plot[{cIV[x,0.5], fn3, fn5}, {x,0,0.9},
     PlotLabel->"Fitted CV Curves - NonU",
     AxesLabel->{"volts","C"},
     PlotStyle->{GrayLevel[0],
     Dashing[{0.01,0.015}],
     Dashing[{0.05,0.02}]}];
```

Out:

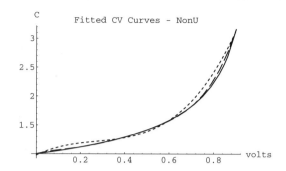

Once we take the following derivatives, we will plot just the responses in the central region and compare those responses to the uniform fitted curves over the same region.

In:

```
dc = D[cIV[x,0.5], x];
dfn3 = D[fn3, x];
dfn5 = D[fn5, x];
Plot[{dc,dfn3,dfn5}, {x,0.2,0.6},
    PlotLabel->"Fitted CV 1stD - NonU",
    AxesLabel->{"volts","dC"},
    PlotStyle->{GrayLevel[0],
    Dashing[{0.01,0.015}], Dashing[{0.05,0.02}]}}];
```

Out:

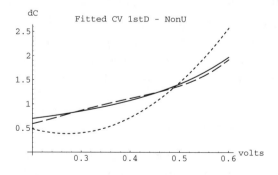

By comparing the nonuniform fit with the magnified uniform fit, as shown below, we can see some improvement in the third-order fit and a great deal of improvement in the derivative of the fifth-order fit.

In:

```
Show[pf1,PlotRange->{{0.2,0.6},{0,2.5}}];
```

Out:

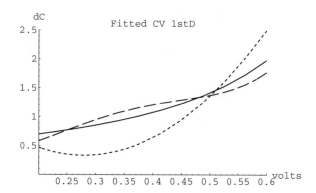

3.3.4 Filtering data to remove noise

Noise in data may make our task of function fitting difficult because noise tends to distract the fitting process – or at least to make it less accurate, especially when the dataset is small. To lessen the effect of noise, we can:

i) take a two-pass approach: we use the first pass to create an approximate fit, and then on the second pass we reject all data that lie farther than some specified distance from the approximate fit;

ii) smooth the data series directly (that is, filter in the time domain);

iii) transfer the data into a domain of basis functions similar to the desired response and filter out basis functions with coefficients below a given level or outside a given eigenvalue.

In all cases, the effectiveness of the noise removal process relies on the existence of some exploitable difference between the noise and the data. If no such difference exists, attempting noise removal may be a waste of time – in which case you will need to seek some other method of improvement, such as investing in a more sensitive detector.

3.3.4.1 Rejecting outliers

Although severe noise in data can result in incorrect fitting, so too can outliers in data. An outlier is a data point with a value that might not be wildly incorrect but which is beyond the typical variations present within its dataset. Outliers can be caused by electrical noise that upsets data acquisition equipment or by a tired researcher who records numbers incorrectly on her notepad.

How you cope with outliers depends on the type of data you are collecting. If you are making many measurements of the same value, then you might decide to discard all those values that have differences in excess of three standard deviations from the mean. If you are recording a timeseries, then the problem is not so easy because you must decide on some criterion that classifies a given value in the data as an outlier.

For example, a surveyor makes 100 measures of some distance (which she stores in the list **distance**) and wants to keep all the measurements that are within one standard deviation of the mean. To find the mean, the standard deviation, and the error of the sample mean, we can use functions in the package **Statistics`DescriptiveStatistics`**:

In:

```
Length[distance]
```

Out:

```
100
```

In:

```
meanDistance=Mean[distance]
```

Out:

```
34.6268
```

In:

```
sd=StandardDeviation[distance]
```

Out:

```
4.51637
```

In:

```
StandardErrorOfSampleMean[distance]
```

Out:

```
0.451637
```

After we have found the mean and the sample standard deviation, we can then use **Select** to pick out and keep only those elements of the original dataset that come within one standard deviation of the mean. You can summarize the output from **Select** by using **Short** (applied here in postfix notation).

In:

```
revisedDistanceData=
    Select[distance,
           Function[Abs[meanDistance-#]<=sd]]//Short
```

Out:

```
{31.6639, 38.9521, 35.1151, <<64>>, 32.882, 36.6025}
```

The size of the sample has now decreased (it now has 69 members), and we have new values for the mean, standard deviation, and error of the mean.

In:

```
Length[revisedDistanceData[[1]]]
```

Out:

```
69
```

In:

```
mean2=Mean[revisedDistanceData[[1]]]
```

Out:

```
35.0434
```

In:

```
sd2=StandardDeviation[revisedDistanceData[[1]]]
```

Out:

```
2.73542
```

In:

```
StandardErrorOfSampleMean[revisedDistanceData[[1]]]
```

Out:

```
0.329305
```

The task of dealing with outliers in cases for which you are measuring a changing quantity (for example, a timeseries) is rather difficult because you have to be able to specify what does (and does not) make some measurement a valid member of the dataset.

For example, if you are measuring the temperature of a block of iron where the sampling interval is much shorter than the thermal timescale of the block, a significantly rogue sample should be easy to spot because the thermal inertia of the block would make such a sudden change impossible. But when the error is of comparable magnitude and/or of similar time-domain character to the noise in your data, there may be no cure.

For now, we limit ourselves to a solution of the iron-block temperature problem. First, we look at the data. Next, we fit an arbitrarily chosen low-order polynomial through the data and plot it and the data to make sure they are valued similarly.

In:

```
ironTempPlot=ListPlot[ironTemp,PlotJoined->False,
    PlotRange->{500,2000},
    AxesLabel->{"time","t(K)"}];
```

Out:

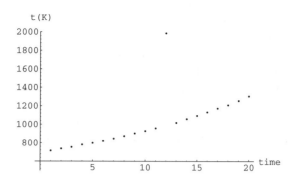

In:

```
ironTempFit=Fit[ironTemp,{1,t^2},t]
```

Out:

$$814.175 + 1.412\ t^2$$

In:

```
Plot[ironTempFit, {t,1,20},DisplayFunction->Identity];
Show[ironTempPlot,%];
```

Out:

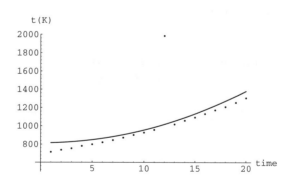

Then we calculate the differences between the data and the polynomial by mapping the fitted function over the data.

In:

```
differences=
Map[Function[{#[[1]],
             Abs[#[[2]]-ironTempFit /. t->#[[1]]]}],
    ironTemp]
```

Out:

```
{{1, 100.339}, {2, 82.3275}, {3, 72.9634},
{4, 55.6874}, {5, 50.2664}, {6, 44.0653}, {7, 38.4913},
{8, 32.4072},{9, 26.6091}, {10, 27.884}, {11, 27.6319},
{12, 967.097}, {13, 35.8927}, {14, 33.0466},
{15, 38.6345}, {16, 44.0764}, {17, 49.4522}, {18,
63.7421}, {19,  69.2559}, {20, 74.3237}}
```

Finally, we determine the element number of the maximum difference value and then drop that element from the dataset.

In:

```
Position[differences,Max[differences]][[1,1]]
```

Out:

```
12
```

In:

```
cleanedData=Drop[ironTemp,{%}]
```

Out:

```
{{1, 715.248}, {2, 737.496}, {3, 753.92}, {4, 781.08},
{5, 799.209}, {6, 820.942}, {7, 844.872}, {8, 872.136},
{9, 901.938}, {10, 927.491}, {11, 957.395}, {13,
1016.91}, {14, 1057.88}, {15, 1093.24}, {16, 1131.57},
{17, 1172.79}, {18, 1207.92}, {19, 1254.65}, {20,
1304.65}}
```

In:

```
ListPlot[cleanedData,PlotJoined->False,
    PlotRange->{500,2000},AxesLabel->{"time","t(K)"}];
```

Out:

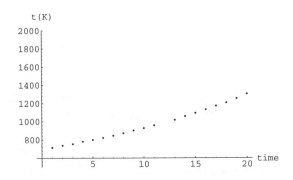

When outliers are not obvious, it is often better to smooth them into the dataset rather than to ignore them.

3.3.4.2 Smoothing and filtering

We now show how the following function, designed by Cornelius Lanczos, can be used to smooth time-series data. This implementation in *Mathematica* was translated from Algorithm 216 of the ACM into *Mathematica*. Many smoothing functions create large errors at the ends of the data series to which they are applied – or else they skew the data series – but the following function, **smooth**, uses central differences to avoid skewing the data and takes special care at the end points of the dataset.

In:

```
smooth::usage =
"smooth[list] returns a 4th-order smoothed set
of numbers according to Lanczos's method in
Applied Analysis. See Algorithm 216 of the ACM due
to R. George.";


smooth[x_] :=
 Module[ {deltaL,factorL, extrL, dataL, nL},
    (* translated to Mathematica by Alfred Riddle *)
    nL = Length[x];
    factorL = 3/35;
    deltaL = Drop[ x,1] - Drop[ x, -1];
    (* make the 4th divided difference &
       keep 3rd end points *)
    Do[(* put list ends in extrL *)
       extrL = Join[Take[deltaL,1], Take[deltaL,-1]];
       deltaL = Drop[ deltaL,1] - Drop[ deltaL, -1];
    ,{3}];
    (* take input, modify ends & join to diffs *)
    dataL = x;
    dataL[[1]] += extrL[[1]]/5 + deltaL[[1]] factorL;
    dataL[[2]] +=  - 0.4 extrL[[1]] - deltaL[[1]]/7;
    dataL[[nL]] += - extrL[[2]]/5 +
                      deltaL[[nL-4]] factorL;
    dataL[[nL-1]] += 0.4 extrL[[2]] - deltaL[[nL-4]]/7;
    dataL -= factorL Join[ {0,0}, deltaL, {0,0}] //N
 ]
```

We will use the dataset from Section 2.7.2 with glitches – which is repeated below and shown again – but this time we pad the dataset with zeros to make it 64 points in length. We are going to look at various ways of smoothing data and Fourier methods are most convenient with sets of data that have lengths equal to a power of 2.

In:

```
dataG = {1,0,0.5,2,3,4,5,6,7,8,9,10,5,
         12,13,14,15,16,17,18,19,20,9.5,
         18,17,16,15,14,13,12,11,10,4.5,8,7,6,5,4,3,2};
dataGpad=Join[dataG,Table[0.0,{64-Length[dataG]}] ];
odPlot = ListPlot[dataGpad, PlotLabel->"Glitchy Data"];
```

Out:

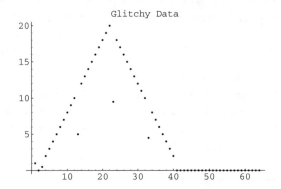

The Fourier transform of this dataset has both large magnitudes at several frequencies, due to the data, and smaller-valued rippling, due partly to the data but mostly to the impulsive noise. We use **Abs** before passing values to the plotting function because the Fourier-transformed values can be complex. (Note the folded-over spectrum – the butterfly effect – in the upper half of the dataset.)

In:

```
ListPlot[Abs[Fourier[dataGpad]], PlotRange->All];
```

Out:

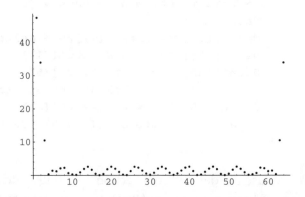

When we filter the glitches using a brick-wall filter – the simplest filter to implement but often not the best for removing noise – we almost always destroy some of the data as well as the glitch noise. The glitches contain mostly high-frequency information, so we select the low-frequency information by filtering to minimize the effect of the glitches. **filterF** is a brick-wall filter for datasets with a folded-over spectrum. The filter works by taking the desired points and adding them to an array of zeros equal in length to the original signal.

In:

```
filterF::usage=
"filterF[signal,point,halfWidth] returns signal with
all points outside of point+/-halfWidth set to zero.
The filter also passes the folded over spectra.";

filterF[sig_,pt_,wd_] :=
  Module[ {tmp,max},
    max = Length[sig];
    tmp = Table[0.0,{max}];
    pts = Join[ Range[pt-wd,pt+wd],
                max + 2 + Range[ -pt - wd, -pt + wd]]];
    (* if filter wider than end of data *)
    pts = Select[pts, (#<=max)&];
    Map[(tmp[[#]] += sig[[#]])&, pts];
    tmp
  ]
```

For this example, the filter is centered at point 2 and has a half-width of 1, so it passes points on either side of the desired main signal.

In:

```
tmp = filterF[Fourier[dataGpad],2,1];
nData = InverseFourier[tmp];
```

The removal of high-frequency information by **filterF** ends up smoothing our response as well as removing the glitches, as shown by **ndPlot**. Transform filtering offers the best chance of removing data errors with minimal distortion of the data – when you know what the data spectrum should look like. Unfortunately, if the frequency content of the data and the errors overlap, then some data will be destroyed while removing the errors.

In:

```
ndPlot = ListPlot[nData,
                  PlotLabel->"Filtered Glitchy Data",
                  PlotStyle->{GrayLevel[0.5]}];
```

Out:

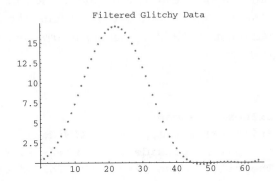

By overlaying the glitchy and filtered data, we can see that the glitches were completely removed. The most significant error in the data is that the rapid transition at the peak of the pulse was removed by the high-frequency filtering.

In:

```
Show[odPlot,ndPlot];
```

Out:

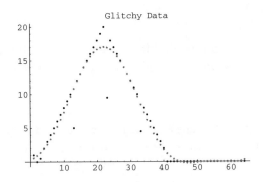

Next we smooth the data and compare that with the filtered data. We use **Nest** to apply the Lanczos smoothing filter to our data several times. (We arbitrarily choose to apply it five times.)

In:

```
dataGs = Nest[smooth,dataGpad,5];
sdPlot = ListPlot[dataGs,
              PlotLabel->"Smoothed Data",
              PlotStyle->{GrayLevel[0.3],
                      PointSize[0.015]}];
```

Out:

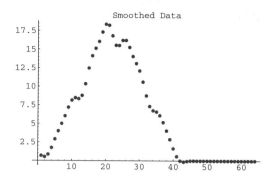

Now the data suffer very little distortion, and the glitches are greatly reduced, but they are not removed as effectively as by transform filtering.

In:

```
Show[odPlot,ndPlot,sdPlot];
```

Out:

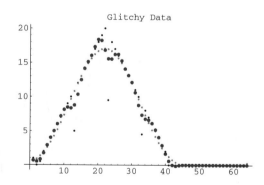

By overlapping the original data, the Fourier-filtered data, and the smoothed data, we are able to see the trade-offs between the different algorithms for smoothing glitches. The Fourier technique preserves the general shape of the curve and removes the glitches best but this techniques also introduces some error at almost all points, including significant error at the peak. The time-smoothed data tend to have errors localized around the glitches; a few smoothings greatly reduce the glitch size.

Another typical error in data is random noise. In this example we will examine a sinusoidal signal with added noise. The original signal, **osPlot**, and the signal with noise, **osnPlot**, are shown below.

In:

```
sig = Table[ Sin[6.28 i/16.0], {i,0,63}];
osPlot = ListPlot[sig,PlotJoined->True];
```

Out:

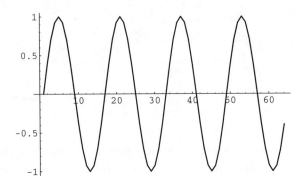

In:

```
noise = Table[Random[],{64}];
osnPlot = ListPlot[sig+noise,PlotJoined->True];
```

Out:

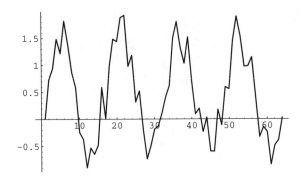

This noise signal also contains a DC offset (because **Random[]** generates random real numbers with a mean of +0.5), like that introduced by the offset voltage of an operational amplifier. Here is the spectrum of the signal and noise.

In:

```
ListPlot[Abs[Fourier[sig+noise]],PlotRange->All,
        PlotLabel->"Spectrum of Signal & Noise",
        AxesLabel->{"Hz","Mag"}];
```

Out:

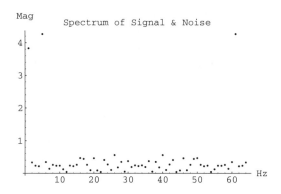

We do not see a periodic pattern in the noise, as we did with the glitchy data; in this case the noise floor is featureless. We filter the signal and noise by sampling the spectrum only at the frequency of the expected signal – in this case, the frequency with the largest magnitude – and the two nearest (adjacent) frequency bins. The reconstructed signal (still with some noise), **nsPlot**, shows some modulation effects from the noise, but the reconstruction does have a clearly defined period with no DC offset.

In:

```
tmp = filterF[Fourier[sig+noise],5,1];
nSig = InverseFourier[tmp];
nsPlot = ListPlot[nSig,PlotJoined->True,
              PlotLabel->"Filtered Signal+Noise",
              PlotStyle->{GrayLevel[0.5]}];
```

Out:

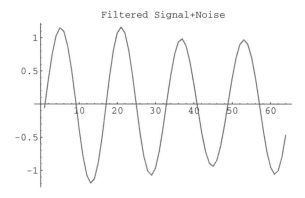

As a comparison, we next smooth the signal and noise with five iterations of **smooth**. The resulting smoothed signal has more noise distortion and the DC offset is not removed. (The offset should not be removed by smoothing, because this smoothing algorithm is equivalent to a low-pass filter.)

In:

```
sigS = sig+noise;
sigS = Nest[smooth,sigS,5];
nSPlot = ListPlot[sigS,PlotJoined->True,
        PlotLabel->"Smoothed Signal+Noise"];
```

Out:

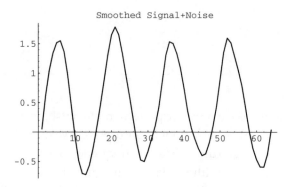

Comparing the results of transform-filtering and smoothing on noise, we notice that the Fourier-filtered signal is very close to the original. Because our transform technique separates the signal into sinusoids, the Fourier technique can best preserve the data.

In:

```
Show[osPlot,nSPlot,nsPlot];
```

Out:

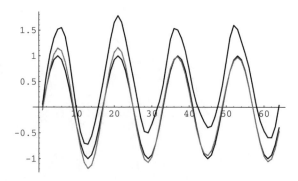

The choice of basis function domain for filtering is important. If we were working with square-wave signals, then Walsh- or Haar-transform filters, discussed by Beauchamp (1975) and Gonzalez & Wintz (1981), would be better suited to noise removal than its Fourier equivalent. Riddle & Dick (1995) also discuss time series manipulation and filtering with *Mathematica*.

3.4 References

Beauchamp, K. G., "Walsh Functions and Their Applications," Academic Press, London, United Kingdom, 1975.

de Boer, C., "A Practical Guide to Splines," Springer-Verlag, New York, USA, 1978.

Gonzalez, R. C., Wintz, P., "Digital Image Processing," Addison-Wesley, Reading, Massachusetts, USA, 1981.

Maas, S. A., "Nonlinear Microwave Circuits," Artech House, Norwood, Massachusetts, USA, 1988.

Michalewicz, Z., "Genetic Algorithms+Data Structures=Evolution Programs (2nd edition)," Springer-Verlag, Berlin and Heidelberg, Germany, 1994.

Pillage, L. T., Rohrer, R. A., Visweswariah, C., "Electronic Circuit and Simulation Methods," McGraw-Hill, New York, USA, 1995.

Press, W. H., Flannery, B. P., Teukolsky, S. A., Vetterling, W. T., "Numerical Recipes," Cambridge University Press, Cambridge, United Kingdom, 1990.

Riddle, A., Dick, S., "Applied Electronic Engineering with *Mathematica*," Addison-Wesley, Reading, Massachusetts, USA, 1995.

Tsui, J. B-Y., "Digital Microwave Receivers," Artech House, Norwood, Massachusetts, USA, 1989.

CHAPTER 4

Generating test data

You need to create test data to help with several tasks in the "hypothesis – test/observation – new hypothesis" loop. For example, you will need to generate test data if you want to try out a new algorithm or program, to compare values generated from a theoretical function with those measured by experiment, or to export values to some equipment – say, a shaker table for testing a product's robustness or colored random sound for evaluating a microphone or loudspeaker. In this chapter we show you how *Mathematica* can generate data, both perfect and noisy.

4.1   Creating perfect data

Many of our examples in the book have already used the function that is simplest to use for generating data: **Table**. There are four main (overloaded) variants of **Table**, and each of them takes a different number of arguments.

In:
```
Table[3.25,{4}]
```
Out:
```
{3.25, 3.25, 3.25, 3.25}
```
In
```
Table[i^2, {i, 3}]
```
Out:
```
{1, 4, 9}
```
In:
```
Table[i^2,{i, 2, 4}]
```
Out:
```
{4, 9, 16}
```
In:
```
Table[i^2,{i, 1, 3, 0.5}]
```

Out:

    {1, 2.25, 4., 6.25, 9.}

There are two other functions that you can use to make table-like objects: **Range**, which works like **Table**, and **Array**, which has fewer modes of use than **Table**.

In:

    Range[4]

Out:

    {1, 2, 3, 4}

In:

    Range[3, 6]

Out:

    {3, 4, 5, 6}

In:

    Range[3, 4, 0.2]

Out:

    {3, 3.2, 3.4, 3.6, 3.8, 4.}

In:

    Array[Sin, 4]

Out:

    {Sin[1], Sin[2], Sin[3], Sin[4]}

In:

    Array[Sin, 4, 0]

Out:

    {0, Sin[1], Sin[2], Sin[3]}

In:

    Array[Sin, 4, 0]//N

Out:

    {0, 0.841471, 0.909297, 0.14112}

Range is also useful for producing exponential-factor ranges.

In:

    10^Range[3]

Out:

    {10, 100, 1000}

You can use one or more of these functions to generate a table of values that are produced by a (theoretical) function. For example, if you want to compare some data to a quadratic function and determine the residuals – the difference between your theoretical function and the empirical data – you might proceed as follows. Note the **\n** at the end of the last heading's text gives us an extra line between the headings and the first line of data. The first argument to the **TableHeadings** option (**None**, in this instance) controls row-describing text.

In:

```
data={1.1, 3.8, 9.2, 16.7, 24.6, 35.9};
function[x_]:=x^2;
TableForm[
 Table[{i,function[i],data[[i]],function[i]-data[[i]]},
      {i,1,6}],
      TableHeadings->{None,
      {"i","f(i)","d(i)","f-d\n"}}]
```

Out:

i	f(i)	d(i)	f-d
1	1	1.1	-0.1
2	4	3.8	0.2
3	9	9.2	-0.2
4	16	16.7	-0.7
5	25	24.6	0.4
6	36	35.9	0.1

Table, **Range**, and **Array** work well when you want to generate evenly spaced values. If you need values that are based on a parameter that has unevenly spaced values, then it is more appropriate to map the function over a list of those values. Using data from the above example, you might want to determine the square root of each value and then to find the difference between the each square root and its corresponding integer.

In:

```
Map[Function[Sqrt[#]], data]
```

Out:

```
{1.04881, 1.94936, 3.03315, 4.08656, 4.95984, 5.99166}
```

In:

```
Range[1,6] - %
```

Out:

```
{-0.0488088, 0.0506411, -0.0331502, -0.0865633,
0.0401613, 0.00833913}
```

4.2   Creating random numbers

Mathematica can generate white or colored random numbers. A white random number is a number that, between two bounds, has an equal probability of occurrence; a fair (unbiased) number cube (die) should produce white random integers between 1 and 6. Colored random numbers have nonflat probability distributions between their bounds.

White random numbers are produced by the function **Random**. You can select both the type of the random number and the bounds. By default, **Random** called with zero arguments will produce a real number between 0.0 and 1.0. You can specify the number type in **Random**'s first argument (which can be **Integer**, **Real**, or **Complex**) and other ranges can be specified as the second argument in a two-element list.

In:

```
Table[Random[], {4}]
```
Out:

```
{0.0327557, 0.413611, 0.741811, 0.126011}
```
In:

```
Table[Random[Integer,{0,10}], {8}]
```
Out:

```
{2, 7, 3, 1, 3, 3, 4, 7}
```

Successive invocations of random will give you different random numbers. If you want the same sequence of random numbers, you need to specify a seed in the function **SeedRandom**, which you need to invoke before each call to **Random**.

In:

```
Table[Random[], {4}]
```
Out:

```
{0.202151, 0.450127, 0.185923, 0.151226}
```
In:

```
SeedRandom[13];
Table[Random[],{4}]
```
Out:

```
{0.103071, 0.801879, 0.170558, 0.194772}
```
In:

```
SeedRandom[13];
Table[Random[],{4}]
```

```
{0.103071, 0.801879, 0.170558, 0.194772}
```

Mathematica can generate random numbers quite quickly. It is always interesting to visualize the distribution of a large set of random numbers, and you can use the **BinCounts** function from the **Statistics** standard packages and **BarChart** from the **Graphics** package to quickly draw out the distribution. (We shall find distribution visualization useful later for looking at colored random numbers.)

In:

```
myRandomData=Table[Random[],{10000}];//Timing
```

Out:

```
{0.816667 Second, Null}
```

In:

```
Needs["Statistics`DataManipulation`"];
Needs["Graphics`Graphics`"];
BarChart[Transpose[{BinCounts[myRandomData, {0,1,0.1}],
                    Table[i,{i,0.05,1,0.1}]}]];
```

Out:

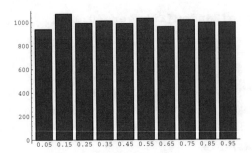

4.2.1 Adding noise to simulated data

Because *Mathematica* treats lists and single-valued symbols consistently, you can add noise to perfect simulated data very easily. The only criterion is that the data list and the noise list must be the same length.

The following example adds noise to a sine wave. Note how we use the **Length** function to ensure that the list of noise is the correct length. Automatically sizing the noise-data list allows you to reuse the same code without having to amend that code to cope with datasets of different length.

In:

```
perfectData=Table[Sin[x],{x,0,2 Pi, Pi/100}];
ListPlot[perfectData,PlotJoined->True];
```

Out:

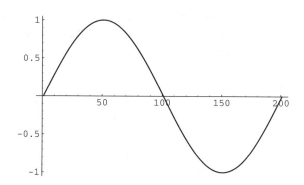

In:

```
noise=Table[Random[]/10,{Length[perfectData]}];
simulatedData=noise+perfectData;
ListPlot[simulatedData,PlotJoined->True];
```

Out:

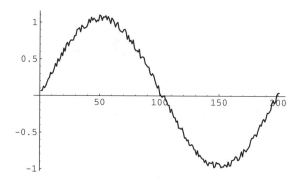

There are very few instances in nature of noise being constrained in value both between two limits and with a flat probability distribution. *Mathematica*'s unmodified **Random[]** function returns values between two limits with a flat distribution; such even-probability noise is often called "white noise." You can generate differently shaped distributions (that is, "colored noise") by specifying the continuous or discrete distribution that *Mathematica* should use when executing **Random**. The specifying functions required are contained in the standard packages **Statistics`ContinuousDistributions`** and **Statistics`DiscreteDistributions`**; we suggest you browse through the documentation for those packages to see what distributions are available. Here is a Gaussian-distribution noise set with a mean of **0.4** and standard deviation of **0.1**. You can use **BinCounts** with **BarChart** or just **ListPlot** to visualize the distribution. (If your computer supports *Mathematica*'s sound functions, remember that you can listen to your random numbers, too.)

In:

```
Needs["Statistics`ContinuousDistributions`"];
Needs["Statistics`DiscreteDistributions`"];
mean=0.4;
sigma=0.1;
myColoredNoise=
    Table[Random[NormalDistribution[mean,sigma]],
        {10000}];
BarChart[Transpose[{BinCounts[myColoredNoise,
                {0,1,0.1}], Table[i,{i,0.05,
        1,0.1}]}], PlotRange->All];
```

Out:

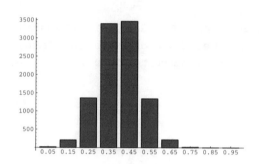

In:

```
ListPlot[myColoredNoise];
```

Out:

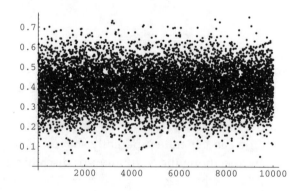

In many areas of the physical sciences, the Poisson distribution is appropriate. For example, a certain amount of a radioactive compound will emit an average of m particles per second and a star of a certain magnitude will deposit a certain number of photons per second onto a detector. Although in both cases the arrival times of particles or photons at the detector is random, there is a well-defined mean arrival rate. By specifying the required distribution to be Poissonian, you can use **Random** to produce numbers that are representative of, say, the number of particles or photons detected in one time interval. For example, if we expect a mean arrival rate of 20 photons per second and we take a large number of 1-second samples, we can plot a histogram that shows the probability of any particular number of photons being counted in 1 second. If you want to visualize how the counts vary second by second, you can plot the list of Poisson-valued random numbers. (We have restricted the range used in plotting the list to show the sample-by-sample variation with more clarity.)

In:

```
mean=20;
myPoissonNoise=
    Table[Random[PoissonDistribution[mean]],
        {10000}];
BarChart[Transpose[{BinCounts[myPoissonNoise,
                {0,75,5}], Table[i,{i,1,75,5}]}],
        PlotRange->All];
```

Out:

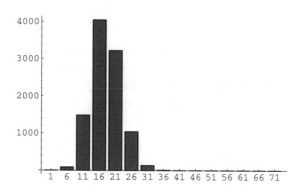

In:

```
ListPlot[myPoissonNoise,
        PlotRange->{{1,100},{0,50}},
        AxesLabel->{"sample","value"}];
```

Out:

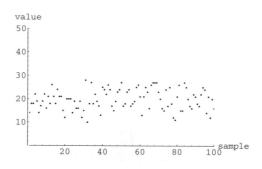

Continuing with Poisson-type data, you can model the arrival times of Poisson random events by noting that the distribution of interarrival gaps is distributed exponentially: the gap between two events is the natural logarithm of a random number (between 0 and 1) divided by m, the mean number of events per sample. We can generate a list of interevent times just by mapping the function **gap** over a list of random numbers:

In:

```
gap[x_,m_]:=-Log[x]/m;
gapData=Map[Function[gap[#,20]],
            Table[Random[],{10000}]];
```

We can check the interevent data by creating a histogram of the number of events counted in 1-second samples; of course, the result is nearly identical in overall shape to the histogram of **myPoissonNoise**, above.

In:

```
elapsedTime=0.0;
sampleLimit=1;
sampleSize=0;
dataIndex=1;
samples={};
While[dataIndex<=Length[gapData],
      While[elapsedTime<sampleLimit &&
dataIndex<=Length[gapData],
            sampleSize++;
            elapsedTime+=gapData[[dataIndex++]];
         ];
      AppendTo[samples,sampleSize];
      sampleLimit++;
      sampleSize=0;
   ];
BarChart[Transpose[{BinCounts[samples, {0,75,5}],
                 Table[i,{i,1,75,5}]}],
         PlotRange->All];
```

Out:

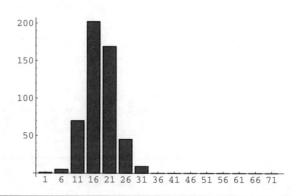

There are two other ways by which we can visualize the qualities of the interevent times: we can plot the times one by one, and we can make a histogram of their occurrence frequency.

In:

```
ListPlot[gapData,
         PlotRange->{{0,100},{0,1}},
         AxesLabel->{"gap\nnumber","gap duration"}];
```

Out:

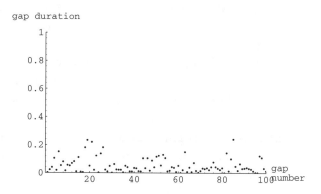

In:

```
BarChart[Transpose[{BinCounts[gapData, {0,0.5,0.1}],
                    Table[i,{i,0.05,0.5,0.1}]}],

         PlotRange->All,
         AxesLabel->{"gap\nduration","occurences"}];
```

Out:

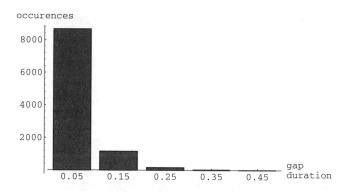

Visualizing simulated data gives you a good feel for the properties of the actual data and can help you decide on parameters to use during an experiment. For example, if you only have a 10-bit counter, what is the maximum resolution that you could obtain while still correctly measuring most of the data? The occurrences plot shows that the probability of a gap greater than 400 ms is quite small (by eye, about 1 part in 10^4), so if the counter can use its 10 bits to count to 400 ms then the best resolution is going to be about 400 μs.

Remember to consider how you will cope when the length of a gap exceeds that handled by the counter without the counter wrapping around to zero. The occurrence histogram, above, gives a good intuitive feeling for how often wrap-around might occur; we can gain a more quantitative measure by using those functions of *Mathematica*'s that return quantities associated with distributions.

For example, the interarrival times for Poissonian events are drawn from the exponential distribution. Above, we used our own function for converting white random numbers into those from an exponential distribution, but we could have chosen to directly modify **Random**'s results by specifying **ExponentialDistribution** as the distribution that **Random** should use. Thus both **gapData** and **myExpData** contain values drawn from the same distribution:

In:

```
myExpData=Table[Random[ExponentialDistribution[20]],
                {10000}];
```

We can now find what fraction of the distribution lies below the value of **0.4** by using the cumulative distribution function **CDF**. We can also determine what value above which 99.99% of the population lies by using the **Quantile** function.

In:

```
CDF[ExponentialDistribution[20],0.4]
```
Out:

```
0.999664537372098
```
In:

```
Quantile[ExponentialDistribution[20],1.0-10.0^-4]
```
Out:

```
0.4605170185988147
```

4.2.2   Random numbers from custom distributions

Sometimes you will want to generate random numbers taken from a distribution that is not a known distribution. For example, if you take data on some process and construct a histogram, you know the distribution for your process. Your process may be the time taken to answer a telephone call or the manufacturing time of a part. These distributions are critical for assessing any service operation. If you run a simulation, naturally you would like to use your actual distribution rather than an assumed (approximate) distribution.

The uniform random number generator in *Mathematica* is transformable into a custom number generator. If $F(x)$ is the cumulative distribution function (CDF) for your custom distribution, and U is a uniformly distributed random variable between 0 and 1, then $X = F^{-1}(U)$.

Your custom random variable is simply a uniformly distributed random number transformed by the inverse of your CDF (see Gordon (1978)). Because the CDF is always between 0 and 1, it naturally reflects a flat distribution into a curve representing your custom distribution without any scaling. The raw data are a set of measured values for telephone response times; each data point is a pair of numbers composed of a time (in minutes) and the number of calls answered between that moment of time and the previous specified time.

In:

```
tResp = {{0,0}, {0.5, 10}, {0.75, 12},{1,23},
{1.5,17},{2,16},{3,12},{5,8},{10,5},{20,3}};
ListPlot[tResp,AxesLabel->{"minutes","#"}];
```

Out:

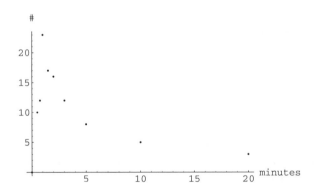

Our CDF for this distribution is given as:

In:

```
{minutes,qtys} = N[Transpose[tResp]];
qCDF = Rest[FoldList[Plus,0,qtys]];
qCDF /= Max[qCDF]
```

Out:

```
{0, 0.0943396, 0.207547, 0.424528, 0.584906, 0.735849,
0.849057, 0.924528, 0.971698, 1.}
```

In:

```
dataCDF = Transpose[{minutes,qCDF}];
ListPlot[dataCDF, AxesLabel->{"minutes","#"},
         PlotJoined->True, PlotLabel->"Call Time CDF"];
```

Out:

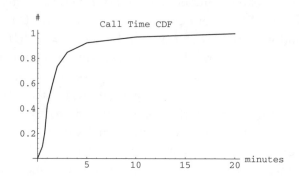

Between any two points, x and y ($= F(x)$) are related by the linear slope: $y_2 - y_1 = a_1 (x_2 - x_1)$, where the 1 and 2 subscripts denote the beginning and ending points on any segment, respectively. Because the a values are just the slopes, we can solve for the slopes from the data.

In:

```
aSlope = (Drop[qCDF,1]-Drop[qCDF,-1])/
            (Drop[minutes,1]-Drop[minutes,-1])
```

Out:

```
{0.188679, 0.45283, 0.867925, 0.320755, 0.301887,
 0.113208, 0.0377358, 0.00943396, 0.00283019}
```

Once we know the segment number for our data point – that is, which bin in the array is appropriate – we can compute our transformed variable by inverting the previous slope equations. We can find the segment number by counting the number of segments with a starting value less than or equal to the desired value. For example, if our uniform random number were 0.7, the segment number would be 5, as given below.

In:

```
Length[Select[qCDF, (#<=0.7)&]]
```

Out:

```
5
```

The inverted discrete CDF is given as follows:

In:

```
Solve[y - y1 == a1 (x - x1), x]
```

Out:

$$\{\{x \to \frac{a1\ x1 + y - y1}{a1}\}\}$$

and the fifth value in each array serves as x_1, a_1, and y_1 respectively. So the corresponding waiting time for a random number valued at 0.7 is:

In:

```
minutes[[5]] + (0.7 - qCDF[[5]])/aSlope[[5]]
```

Out:

```
1.88125
```

Now we can build a distribution generator function by collecting all the previous steps.

In:

```
CustomGenerator[cdfVals_List,rv_] :=
  Module[ {xs,ys,as,pt},
          {xs,ys} = Transpose[cdfVals];
          as = (Drop[ys,1]-Drop[ys,-1])/
                 (Drop[xs,1]-Drop[xs,-1]);
          pt = Length[Select[ys,(#<=rv)&]];
          xs[[pt]] + (rv-ys[[pt]])/as[[pt]]
        ]
```

As a check, we can rerun our initial analysis point of 0.7.

In:

```
CustomGenerator[dataCDF,0.7]
```

Out:

```
1.88125
```

We can also plot the inverse distribution.

In:

```
Plot[CustomGenerator[dataCDF,x],{x,0.01,0.99},
PlotLabel->"Inverse Custom Distribution"];
```

Out:

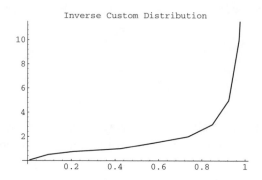

The real test, however, is to create our desired distribution from random numbers compliant with the inverse distribution.

In:

```
u = Table[Random[], {1000}];
xInv = CustomGenerator[dataCDF,#]& /@ u;
Needs["Statistics`DataManipulation`"]
bc = BinCounts[ xInv, {0,10,0.5} ]
```

Out:

```
{104, 315, 158, 147, 61, 65, 19, 18, 27, 14, 3, 4, 5,
6, 3, 5, 2, 7, 7, 4}
```

To analyze our constructed distribution, we plot the bin values with the corresponding bin counts by creating a list with **Range** and **Rest**. (**Rest** is used because the number of bins is one fewer than the number of data divisions.)

In:

```
Needs["Graphics`Graphics`"]
BarChart[ Transpose[{bc,Rest[Range[0,10,0.5]]}] ];
```

Out:

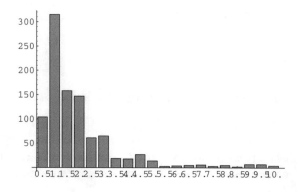

By itself, the function **BarChart** has crowded together all the bar labels. If we take control of generating the labels, we can create a bar chart that is much easier to comprehend.

In:

```
lablList=Table[If[IntegerQ[labl],ToString[labl],""],
              {labl,1/2,10+1/2,1/2}]
```

Out:

```
{, 1, , 2, , 3, , 4, , 5, , 6, , 7, , 8, , 9, , 10, }
```

In:

```
BarChart[ Transpose[{bc, Rest[Range [0,10,0.5]]}],
PlotRange->All, BarLabels->lablList ];
```

Out:

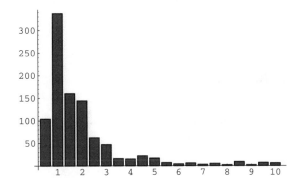

4.3   Reference

Gordon, G., "System Simulation," Prentice-Hall, Englewood Cliffs, New Jersey, USA, 1978.

CHAPTER 5

Exporting data

Mathematica can export either ASCII or binary data to files either by creating new files or by appending to files that already exist. In this chapter, we show you how to use the built-in functions to export integers, real numbers, and strings, all in a variety of formats.

5.1   Basic file operations

To choose the default directory for output, you can use **SetDirectory**, just as we did for importing data.

In:

```
SetDirectory["Q127Beta:Sam's folder"]
```

Remember to use the correct physical volume/disk and directory-level separators for your computer's operating system. (DOS systems use a backslash which is **** in *Mathematica*; **** is *Mathematica*'s line continuation character.) Other related file handling functions include **FileNames[]**, which returns a list of files in the current directory; **FileNames["myFile.*"]** which returns the names of files in the current directory that match the template **myFile.***, where * is the wildcard character; **Directory[]** which returns the path of the current directory; **DeleteFile["name"]**, which deletes the named file; and **CopyFile["f1","f2"]**, which copies file **f1** to a file named **f2**.

5.2   *Mathematica* output to ASCII files

Mathematica has sufficient exporting functions to make writing either *Mathematica* expressions or other ASCII data to files quite simple.

5.2.1 Evaluated and unevaluated expressions

Exporting either a *Mathematica* expression or data created or manipulated by *Mathematica* to a file is simple, and the infix format of the function mimics the Unix shell operator **>** and the C++ iostream operator **>>**. For example,

In:

```
Put[ Table[i,{i,1,10}], "outFile" ]
```

is equivalent to

In:

```
Table[i,{i,1,10}] >> "outFile"
```

Both will store the evaluated expression to the left of the **Put** (**>>**) operator in the file called **"outFile"**. If you want to append output from *Mathematica* onto the end of a file, then you need to use the **PutAppend** (**>>>**) operator. Using the long form of **PutAppend**,

In:

```
PutAppend[Pi, "outFile"]
```

will append **Pi** onto the end of **outFile**. If you examine the contents of **outFile** you will see that *Mathematica* has evaluated the table but has left **Pi** unevaluated because this is the most precise notation for π. If you did not want the evaluated format to be exported, then wrap the function **HoldForm** around the expression you want to export.

In:

```
Put[HoldForm[Table[i,{i,1,10}]], "outFile2"]
```

This will export the expression, complete with the **HoldForm** wrapper, so that if the expression is imported back into *Mathematica*, it will be held unevaluated.

5.2.2 Definitions

To successfully export function definitions, you need to use a different mechanism. For example, if you define a function and then use **Put** to store the function name in a file **eek**, only the name will be stored, not the body of the function.

In:

```
myFunction[x_]:=x+2;
myFunction >> "eek"
```

To store the complete definition, you need to use the **Save** function. **Save** takes two or more arguments: the first argument is the destination filename and subsequent arguments are the functions whose definitions are to be saved.

In:

```
Save["eek2", myFunction]
```

Like **Put** and **PutAppend**, **Save** stores information in ASCII form so that you can type or print the file or import it as ASCII text into a word-processing program. (The inverse function to **Put** is **Get**.)

5.2.3 Expressions in C/FORTRAN/TeX forms

You can use *Mathematica* to format output into a more useful form. For example, if you have used *Mathematica* to generate a symbolic result that you then want to incorporate into your C or FORTRAN program, you can use the format-modifying functions **CForm** and **FortranForm** to write the program code for you. Not only does this save you time, it also helps to eliminate mistakes in interpretation or copying.

In:

```
oneOverXPlusTwo[x_]:=1/(x+2)^2
nested=Nest[oneOverXPlusTwo,z,3]
```

Out 2.2:

```
                  -2 -2 -2
(2 + (2 + (2 + z)   )   )
```

In:

```
CForm[nested]
```

Out:

```
Power(2 + Power(2 + Power(2 + z,-2),-2),-2)
```

In:

```
FortranForm[nested]
```

Out:

```
(2 + (2 + (2 + z)**(-2))**(-2))**(-2)
```

If you are preparing a scientific paper or report, you may want to convert a *Mathematica* result into TeX so that you can import it into your TeX or LaTeX document:

In:

```
TeXForm[nested]
```

Out:

```
{{\left( 2 + {{\left( 2 + {{\left( 2 + z
            \right) }^{-2}} \right) }^{-2}}
            \right) }^{-2}}
```

5.2.4 Formatted numbers

If you want to export numbers into files that will subsequently be read by existing applications, those applications may require that the numbers be formatted in a very particular way. Such a requirement is especially true of aged FORTRAN programs with formatted reads; note that Shaw & Tigg (1994) give a particularly good treatment of both reading and writing FOR-TRAN formatted numbers. Before we look at how you can produce tightly formatted output, let us look at free format.

Where loose control is allowable over the format, the three base versions of the ***Form** functions can be used. For example, the function **ScientificForm**[n] expresses n as a number from 0 to <10 multiplied by 10 raised to a power; **EngineeringForm** constrains the power of 10 to be a multiple of 3. In comparison, **AccountingForm** never uses scientific notation – and displays negative numbers in brackets, typical of accounting habit.

In:

```
n=-0.000342;
ScientificForm[n]
```
Out:

```
        -4
-3.42 10
```
In:

```
EngineeringForm[n]
```
Out:

```
      -6
-342. 10
```
In:

```
AccountingForm[n]
```
Out:

```
(0.000342)
```

To export numbers as *Mathematica* outputs them, you might use:

In:

```
outStream=OpenWrite["myOutputTestFile"];
Write[outStream, EngineeringForm[n]];
Close[outStream];
```

If you examine **myNumberFile**, you will see that *Mathematica* has left the number within the wrapper **EngineeringForm[]** – not a very useful form for reading by, say, a C or Pascal program. This lack of formatting is because the **EngineeringForm** function alters its argument(s) only at print time, not when any other operation is attempted. If you want to write out the numbers in traditional scientific free format, **FortranForm** or **CForm** are better choices:

In:

```
outStream=OpenWrite["myOutputTestFile2"];
nn=-0.0000000456;
space=" "//OutputForm;
Write[outStream,
        FortranForm[nn], space, CForm[n]]
Close[outStream];
```

Where you want tighter control over formatting, you have to exercise rather more care. You have to be prepared to put aside any formatting techniques that you have learned from other languages. The functions that you will use to open and close the output file remain the same.

In:

```
outStream=OpenWrite["myOutputTestFile3"];
```

The function **PaddedForm** is very flexible but works in slightly different ways to the formatting facilities in FORTRAN or C. Its first argument is the number that you want formatted; the second argument is a two-element list that specifies the total number of digits (excluding the sign and the decimal point) and the maximum number of digits that can be printed as the fractional part (that is, the maximum number of decimal places).

In:

```
aa=PaddedForm[1.23,{6,4},
                NumberPadding->{"z","x"},
                NumberSeparator->"",
                NumberSigns->{"-","+"},
                SignPadding->True]//OutputForm
```

Out:

```
+z1.23xx
```

In:

```
ab=PaddedForm[1.23,{4,5},
                NumberPadding->{"z","x"},
                NumberSeparator->"",
                NumberSigns->{"-","p"},
                SignPadding->True]//OutputForm
```

Out:

```
p1.23xxx
```

In:

```
Write[outStream,
        aa,
        space,
        CForm[2.3],
        ab]
Close[outStream];
```

Note that if you get the format wrong and try to fit too large a number into the format, whereas FORTRAN would print a group of asterisks, *Mathematica* tries to cope. For example, if you were expecting the format code {4, 3} to produce the string .23x (where the x is a padding character), then your data file would have been corrupted.

In:

```
ab=PaddedForm[.23,{4,3},
                NumberPadding->{"z","x"},
                NumberSeparator->"",
                NumberSigns->{"-","+"},
                SignPadding->True]//OutputForm
```

Out:

```
+0.23x
```

To check how a particular formatting scheme will work, you can always map your formatting function over a list of typical numbers.

In:

```
myList={10^-10,0.0001,0.1,1,1000,10^6}//N;
myTestList=Flatten[AppendTo[myList,-myList]];

Map[NumberForm[#,{6,2},
                NumberPadding->{"z","x"},
                NumberSeparator->"",
                NumberSigns->{"-","+"},
                SignPadding->True]&,
        myTestList]
```

Out:

```
                 -10
{+zzz1.xx 10    , +zzz0.xx, +zzz0.1x, +zzz1.xx,
                       6                 -10
 +1000.xx, +zzz1.xx 10 , -zzz1.xx 10    , -zzz0.xx,
                                              6
 -zzz0.1x, -zzz1.xx, -1000.xx, -zzz1.xx 10 }
```

The formatting of **myList** has highlighted the problem of scientific notation: sometimes you do not want numbers expressed with an exponent. You can control how *Mathematica* breaks into scientific notation by specifying the **ExponentFunction** option in **NumberForm**. What you need to supply to **ExponentFunction** is an anonymous function that takes as its single argument the number that would be the exponent. For example, a function **Function[If[-8<#<8,Null,#]]** returns the exponent as **Null** if it lies between **+8** and **-8**. Numbers with an exponent in that range are printed in normal notation; outside that range, scientific notation is used.

In:

```
Map[NumberForm[#,{6,2},
            NumberPadding->{"z","x"},
            NumberSeparator->"",
            NumberSigns->{"-","+"},
            SignPadding->True,
            ExponentFunction->(Function[If[-
8<#<8,Null,#]])]&,
    myTestList]
```

Out:

```
                 -10
{+zzz1.xx 10    , +zzz0.xx, +zzz0.1x, +zzz1.xx,
                                         -10
 +1000.xx, +1000000.xx, -zzz1.xx 10    , -zzz0.xx,
 -zzz0.1x, -zzz1.xx, -1000.xx, -1000000.xx}
```

5.3 *Mathematica* output to binary files

The default output format from *Mathematica* is ASCII text. Even *Mathematica* Notebooks are stored as ASCII files, including all the formatting information. Without doubt, ASCII text format is the file format most easily transferred between programs, even if these programs execute on different computers or operating systems. In general, it is always easiest to leave a space or some other unique character between numerical values so that they can be read in using free format reads. Space-separated entries are also easier to read by eye.

Put and **PutAppend** cope well with exporting ASCII data (we discuss the export of formatted numbers below). If you want to export data to binary files, perhaps to save disk space, then you need to use functions in the standard package **Utilities`BinaryFiles.m`**. (Before you load any standard packages, if you have been using **SetDirectory**, you should check that the default directory or the search path, **$Path**, is set to that containing the *Mathematica* packages.)

In:

```
Needs["Utilities`BinaryFiles`"];
```

After you have loaded the binary file – handling package, the next task is to open the file and obtain a stream identifier, which we assign to the variable **openStream**. Note that the result of **OpenWriteBinary** will be different on different computer systems, as well as between invocations within the same *Mathematica* session.

In:

```
outStream=OpenWriteBinary["myBinaryFile"]
```
Out:

```
OutputStream[myBinaryFile, 12]
```

5.3.1   Numbers

Before we write a number to a binary file, we need to consider its type and format. Numbers can be integers or real numbers.

Integers can be signed or unsigned and 1, 2, or 4 bytes in length (that is, 8, 16, or 32 bits), so we need to specify the type of integer representation that the export process will use. The binary file package supports **Int8**, **Int16**, and **Int32** types for unsigned integers and **SignedInt8**, **SignedInt16**, and **SignedInt32** for signed integers. The type **Int8** copes with numbers from 0 to 255 and the type **SignedInt8** with numbers ranging from -128 to $+127$.

If you are going to read in exported numbers in C, then the 8-, 16-, and 32-bit length integers correspond to the C/C++ types **char**, **short int**, and **long int**; the unsigned version of integers in C/C++ have the word unsigned in front of the type (for example, **unsigned long int**). In FOR-TRAN, the corresponding declarations would be **BYTE**, **INTEGER*16**, and **INTEGER*32**. Most Pascal compilers use **INTEGER** and **LONGINT** for 16- and 32-bit integers respectively.

Real (or floating-point) numbers can be represented in export with 4 or 8 bytes, with both lengths compliant with the IEEE standard. The *Mathematica* type names for these are **Single** and **Double**. In C/C++ these floating point types correspond to **float** and **short double** on most compilers; the type **double** often uses the native length of the computer's floating-point hardware processor and may be 8, 10, or 12 bytes, typically. In FORTRAN, the corresponding declarations would be **REAL*4** and **REAL*8**. Some Pascal compilers use the names **REAL** or **SINGLE** and **LONGREAL**.

Last but not least, you need to know whether your computer is a little-endian or a big-endian. These strange but often used terms refer to the order that a computer stores the bytes in a multi-byte number. The bytes in a number are stored sequentially in memory, so it is possible to start with either the least or the most significant byte, and thus the terms little-endian and big-endian, respectively. In the function **ToBytes**, you need to specify **LeastSignificantByteFirst** or **MostSignificantByteFirst** for the option **ByteOrder->**.

We strongly recommend checking your compiler's documentation for information on how it treats numbers of different formats and on how it complies with the IEEE floating-point number standard.

After all that, exporting an integer and a floating-point real number is not too difficult:

In:

```
i=16;
WriteBinary[outStream,
            i,
            ByteConversion->
                (ToBytes[#,Byte]&)];


r=16.456;
WriteBinary[outStream,
            r,
            ByteConversion->
                (ToBytes[#,
                        RealConvert->Single]&)];
```

Once you have finished writing to a file, do not forget to close it. Closing the file forces the operating system to complete any pending writes and to tidily close the file.

In:

```
Close[outStream]
```

Out:

```
myBinaryFile
```

5.3.2 Strings

Exporting a string is somewhat easier. A *Mathematica* string is converted to a C-type string, exported as a sequence of two-byte integers, the last of which is zero-valued to indicate the end of the string. The following code will export the string **s** to **outStream**:

In:

```
s="data";
WriteBinary[outStream,
            ToBytes[s,
                    StringConvert -> CString]]
```

You can easily convert strings into other formats. For example, you might want to export a string as a true C-type string or a Pascal-type string. To convert a *Mathematica* string into a form suitable for export as a C-type string, you first convert its characters into their ASCII values and then append the C-string's terminating null (0) character onto the end of the list of ASCII values. It is very important to make sure the string terminates! The functions **OpenWriteBinary** and **Close** work as before. The **WriteBinary** function takes three arguments: the output stream identifier, the list of bytes to be written, and a function that specifies how the elements of that list are to be converted. Because the ASCII values of the string characters translate directly into byte values, the function simply takes the value passed to it and passes it on, unaltered. The pure function that does this passing-on returns its argument without modification.

In:

```
sC="my C text string";
sCList=ToCharacterCode[sC]
```

Out:

```
{109, 121, 32, 67, 32, 116, 101, 120, 116, 32, 115,
116, 114, 105, 110, 103}
```

In:

```
AppendTo[sCList,0]
```

Out:

```
{109, 121, 32, 67, 32, 116, 101, 120, 116, 32, 115,
116, 114, 105, 110, 103, 0}
```

In:

```
outStream=OpenWriteBinary["CStringFile"]
OutputStream[CStringFile, 32]
WriteBinary[outStream,
            sCList,
```

```
                    ByteConversion->Function[#]]
    Close[outStream]
```

Out:

```
    CStringFile
```

In Pascal, strings are contained in a list of bytes with the first byte in the list containing the number of characters (bytes) in the string. So the string **"at"** would be represented by the three-element list of bytes **{2, 97, 116}**. Of course, Pascal strings must be shorter than 255 characters, or else the string length will overflow the byte-long length holder. This example works similarly to the C-type string example except that no null terminator is required. Instead, the string length is prepended to the front of the list.

In:

```
    sPascal="my Pascal string";
    sPascalList=ToCharacterCode[sPascal]
```

Out:

```
    {109, 121, 32, 80, 97, 115, 99, 97, 108, 32, 115, 116,
    114, 105, 110, 103}
```

In:

```
    PrependTo[sPascalList,Length[sPascalList]]
```

Out:

```
    {16, 109, 121, 32, 80, 97, 115, 99, 97, 108, 32, 115,
    116, 114, 105, 110, 103}
```

In:

```
    outStream=OpenWriteBinary["PascalStringFile"]
    OutputStream[PascalStringFile, 27]
    WriteBinary[outStream,
                sPascalList,
                ByteConversion->Function[#]]
    Close[outStream]
```

Out:

```
    PascalStringFile
```

5.4   Reference

Shaw, W. T., Tigg, J., "Applied *Mathematica*: Getting Started, Getting It Done," Addison-Wesley, Reading, Massachusetts, USA, 1994.

CHAPTER 6

Introduction to instrument control and data acquisition

In the remaining chapters of this book, we describe how you can use *Mathematica* to control instruments in your laboratory. Until now, we have discussed software that operated on files on your computer's disk store or that manipulated data generated within a *Mathematica* session. But most scientific and engineering experiments use external equipment to gather data. For example, your data may come from a multi-channel voltmeter that sits on your workbench. That voltmeter may be controlled "stand-alone" by pressing buttons on its front panel, or it may be operated "remotely" by issuing commands from some other equipment or a computer.

If the meter operates stand-alone, you probably have to write your measurements on paper and then type them into your computer. If you have a remotely controlled meter, you are probably using a program to set up the meter (for example, to adjust its input range and sampling frequency) and to acquire the data into a file. Further interaction with your data may mean that you have to transfer the data into another program – perhaps a spreadsheet, some code you wrote in C or another programming language, or *Mathematica*.

Would it save you time to be able to set up and to acquire and analyze your data from within *Mathematica*? This question is not simple to answer because, like all software, *Mathematica* has its strong points and its weak points. It also has to work with some code that lies between it and the equipment that you want to control.

In this chapter, to help you choose how to tackle controlling your laboratory instrumentation from *Mathematica*, we discuss some of the design issues, the typical architecture of a data acquisition system, and some of the other issues that you may encounter when you consider how to store or transmit your data.

6.1   Anatomy of an instrument data acquisition system

One of our favorite drawings is in the book *Object-oriented Analysis and Design* by Booch (1994). In this drawing, a cat is being observed by two people, the cat owner and a veterinarian. The owner sees the cat as a furry friend who likes food, purrs, and adores the owner; the veterinarian sees the same cat as a collection of functional organs, a skeleton, and an epidermis. When we look at data acquisition systems, we need to look at them in different ways, too. In this section we look at two aspects of systems: how their compliance with common standards simplifies your work, and how you might tackle the design of your software.

6.1.1   Standards in instrumentation

If you are somewhat anxious about approaching the task of instrument control, then we hope this section will help dispel some of your worries. Take comfort, too, from remembering that regardless of how complicated any instrument is, its makers do want you to be able to use it!

Historically, if you wanted to control an instrument from your computer, you probably would have found it necessary to write some of the low-level code that drives the instrument and interacts with the computer's operating system. Such code probably would have been in assembly language or, even if it was in a higher-level language like FORTRAN, would still have had to work with the instrument at a bit-setting level.

Today, hardware interface cards come with full-functionality libraries that you can call from high-level languages; you should not have to write bit-setting code or use assembly language. Typically, manufacturers provide C-, FORTRAN-, and sometimes Pascal-callable libraries. For IBM-compatible and Apple Macintosh computers – the main two machine types that have a rich variety of tightly costed interface cards – the subroutine and function libraries supplied with your card will support and be supported by the most common code development systems, including a combined compiler, symbolic debugger, linker, and executable-image maker, and also usually code-management/version-management functionality. Verifying that your compiler is supported (that is, that the supplied libraries and other files can be compiled by and/or linked with that code development system) will save you work later.

From a system-wide viewpoint, any instrumentation system will consist of an instrument, an interface card, a computer, a library of functions or subroutines, a code development package, and *Mathematica*. In some cases, the instrument itself will be on the interface card; cards with analog signal measuring functions, for example, are commonly available. In other cases,

Figure 6a   The components of a data acquisition system

the instrument must be addressed over some kind of communications link such as an RS-232 serial link or an IEEE-488 parallel link.

You should determine what actions you require your instrument to do, what commands you need to issue from the interface card to carry out those functions, and what you want to do with the data once you import them into *Mathematica*. You should also find out from the library documentation what header files (.h in C) that your software requires and what libraries you need to link with your program's object code to make the executable image. Most suppliers provide a few examples to help you gain confidence as you take your first steps.

Although this overview might be simple, it will be applicable to most data acquisition systems. For example, suppose you want to reset your digital voltmeter. If the voltmeter is

- on the interface card itself, then you will need to call the function that resets the meter.
- on a communications link, then you will need to invoke the function that will send the reset command over the link.

In the first case, once you have software that successfully interacts with the instrument card for one function, using more of the card's functions should be straightforward. In the second case, once you have mastered sending and receiving a simple message over the communications link, more complicated operations should seem more tractable. Two common interface standards are either the RS-232, or RS-422/485 (serial) and the IEEE-488 (parallel). Instruments on cards that you plug directly into the computer's bus or backplane are controlled in a similar manner, regardless of the specific bus type. In Chapter 9, we give a NuBus example, but the principles involved would apply to instruments with libraries for other bus standards, including EISA, PCI, and VME.

Controlling any instrument should involve no more work than calling a **TAN** or **SQRT** function. But, just as you need to know what scientific functions are supplied with your code development system before you write some statistical analysis or other mathematically based code, so do you need to know what functions are available for you to call to control your instrument.

6.1.2   Designing your software

Just as we have looked, albeit quickly, at the hardware architecture of a system, we need to look (in somewhat greater depth) at the system's software. The software can be analyzed at different specified "layers," or from different hierarchical perspectives. In this way the functionality of the software can be broken down into various depths of detail.

These abstract layers are grasped more easily when each is attached to a practical purpose. For example, a biology researcher wants to know how the growth of tomatoes is affected by the daily cycle of temperature variation in a greenhouse. How can we transform this problem into layers, and how does our transformation influence the design of the software?

Traditionally, the top layer is the highest level description of the problem in the language of the user. Thus the bottom layer may involve the specification of engineering detail, such as the type of electrical connectors to be used, or how long the battery will last when the equipment is deployed in the field. Although there is a clear hierarchy in this context of detail and purpose, not every hierarchy is implementable, nor is the flow of design necessarily one-way. Each layer must be capable of executing the expectations of the layer above, while at the same time having its own expectations met by the layers below. If such capability is not present, the whole hierarchy is flawed and must be redesigned.

Paradoxically, the lowest layer may actually dictate the functionality of the top layer. Thus, neither a top-down nor a bottom-up design approach can be used every time. Sometimes, you have to start at both ends and work inwards, adjusting the parameters of the problem as you go along. The key to a successful design is to remain as flexible as possible while sticking firmly to the aim of achieving the performance goal for the system.

For small projects – typical of many scientific laboratory experiments – it should be possible for you to specify, visualize, check, and implement all of the layers. For large projects, however, it may not be possible for one person to understand all the details in each layer. In these situations, then, where different people have responsibility for different layers, good personal communication within the project becomes vital (Shlaer & Mellor (1988)).

6.1.3 Layerology

In this section we discuss the layers as we see them. Of course, our interpretations of the application problem and its layer boundaries are not uniquely correct, nor is the order of the layers necessarily optimal; you may have a better collection of layers.

6.1.3.1 Top (application) layer

Returning to our greenhouse example, the application layer poses the question, "How is the growth of tomatoes affected by the daily cycle of temperature variation in a greenhouse?" To answer this question, we will need to measure both the temperature and the plant growth using some equipment. (We discuss only the temperature measurement here.) It is also reasonable for the application layer to specify any other parameters related to the desired data. For example, the answers to the following questions may affect the scientific usefulness of any data that are gathered.

- How often does the temperature need to be measured? (Hourly?)
- What accuracy is required? (To $\pm 0.50\,°C$?)
- What precision is necessary? ($0.1°C$, $0.001\,°C$?)
- What temperature range will be measured? ($-10\,°C$ to $+50\,°C$?)
- Is one measurement site representative of the entire greenhouse?
- How long will the experiment last?
- What sort of disruptions can be tolerated in the greenhouse?
- Do the measurements need to be available immediately after they are taken?
- How can the measurements be assessed for quality?

Although this is not an exhaustive list, there are quite a number of issues that you need to address within the application layer. You should expect to revisit some of your initial decisions after you have explored the other layers.

6.1.3.2 Hardware assessment layer

Many system designers would consider the hardware at the bottom of the layer hierarchy, and so usually it would come last in any study of the system. For this experiment, however, we choose to create a hardware assessment layer because any restrictions encountered here will need to be fed back into the application layer with some urgency. For example, if the application layer requires a temperature measurement resolution of $0.001\,°C$ but your only available sensor has a resolution of $0.05\,°C$, what are you going to do? Unless you alter some of the application layer requirements or persuade the laboratory director to release more funds with which you can buy a better sensor, there may be little point in carrying out the design of the middle layers, let alone the experiment itself.

There are many (often very different) parameters to be assessed in this layer. For example, if the temperature measurement will be carried out using a sensor and a voltmeter, you need to consider whether the impedance of the voltmeter will burden the sensor's output, how well the meter's measuring range corresponds to the output voltage of the sensor over the expected temperature range, and how well the sensor and meter accuracy and the meter precision match the expectations of the application layer.

In this layer, you should also assess such parameters as the physical volume and weight requirements, the electrical power consumption required, whether the exhaust heat from the equipment will affect the measurements, the suitability and safety of the physical environment into which the equipment has to be placed, what electrical cables will be needed, what plugs and sockets need to be fitted to each cable and item of equipment, and any dangers that may be difficult to control, such as water ingress or other damage causes (such as being stepped upon!).

6.1.3.3 The computing layer

If, as we hope, the meter is computer-readable, how will you read the meter? Does your computer require an interface card, or can it use the computer's built-in serial port as a link to the meter? If an interface card is required, you will need to locate the software and the manual that were supplied with the card so that you will be able to drive it.

When you design this layer, there are other bits of information to assemble. You will find it helpful to make, or to have available, a list of all the meter's computer-commandable functions. You should also establish

- how the meter formats numbers (as a string, as binary, as BCD,...)
- what state the meter adopts after power-on and power-fail
- what modes can be selected simultaneously
- whether commands are acknowledged
- how long the meter takes to respond to commands
- how data are transferred (byte by byte or in blocks)
- how you know when a data or command transfer has finished
- what happens if some data (in a block) are missing
- what happens if the computer crashes
- what happens if the computer's receive buffer fills up.

The answers to all these questions should be in the manufacturer's manual for the equipment.

There are other considerations about how functionality is distributed throughout the system. Within a *MathLink* data acquisition system there are three main sites where any activity can happen: in *Mathematica*, in your code that forms the *MathLink* extension to *Mathematica*, or in the hardware itself. For example, if you want to take 200 measurements in a short time, three possible options exist:

i) You could execute a *Mathematica* function 200 times. Each time that function is called, it calls a *MathLink* function that reads one value from the equipment.

ii) You could execute a *Mathematica* function once. Upon execution, that function causes the *MathLink* function to read 200 values from the equipment, one at a time.

iii) If the equipment has a data logger, you might be able to program it with one *Mathematica* function and one *MathLink* function to take 200 samples.

Which method you choose will depend on the in-built functions of the equipment and on the duration of the experiment. For quick series of samples, for example, method (*ii*) might prove to be faster. We urge you to experiment with timings to see what is best for your application. Often it is much easier and more definitive to answer data acquisition questions by performing practical tests rather than by theorizing.

Another point to consider is whether it is simpler to cope with errors returned by the equipment in the *MathLink* code or in *Mathematica*.

Lastly, instrument control is an ideal problem for object-oriented design. All instrument have both a fixed set of instructions to which they respond and a set of data members. Although a language that supports objects is useful, *MathLink* works just fine with C.

The topic of object-oriented design now has a wide literature; Booch (1994), Sexton (1993), and Shlaer & Mellor (1988) are, amongst others, concise and practical guides to the subject.

6.2 Coping with your data

The acquisition of data may not be the only topic that you need to consider when you design your data acquisition system. Where large volumes of data are present, you should consider how long data handling (such as transmission, assessment, or reduction) will take. By careful design of the data handling method, you may be able to significantly shorten the time it takes you to complete your experiment.

6.2.1 Data assessment

How are you going to know if the data you are acquiring are valid?

Assuming there are some criteria by which you can judge data to be acceptable, you can apply those criteria in either your C-based *MathLink* code or your *Mathematica* session. Checking in the *MathLink* code will probably allow you to spot problems with the data acquisition process sooner than if you wait until the faulty data are transmitted into your *Mathematica* session; nevertheless, it might be easier to check your data in *Mathematica*.

For example, you might want to check the mode or median value of the collected data – both are good representatives of the typical acquired value, since the mean can be biased by sporadic large values that might be caused by occasional instrument problems. If you choose to have such checking in your *MathLink* code, that code must sort the acquired data in order to calculate the median. Sort routines are commonly available in published collections of algorithms such as C Users' Group (1994) or Press et al. (1994) or in computer science texts, such as Horowitz & Sahni (1982). So, you should not have to reinvent a sort algorithm, although you do need to accurately implement the code.

On the other hand, checking the mode or median after you have the data up loaded into *Mathematica* is somewhat simpler: The functions **Mode** and **Median** are in the **Statistics `DescriptiveStatistics`** package. *Mathematica*'s functions for drawing graphs are also available to help you visualize your data.

If it is possible for you to collect a small subset of the data before proceeding to collect the principal bulk of the data, then most likely it will be easier to analyze and visualize the data in *Mathematica*. As a very rough guide, you might expect *Mathematica* to take approximately 20 seconds on a typical 60 MHz desktop personal computer or workstation to plot a 10,000-point graph. (Such timings may vary significantly between machines, depending on the *Mathematica* memory configuration, the computer's use of virtual memory, and so on.) To quickly assess large volumes of data, you might want to use a simple value-checking routine in your *MathLink* code and then use *Mathematica* to plot summary data – for example, the median, minimum, maximum, and mean of the data.

6.2.2 Transmission time

Where large volumes of data are being transferred, you might want to consider the time required for these operations.

Normally, serial standards are slower than parallel standards, although there are exceptions. For example, an RS-232 serial link to a small sensor might run at 9600 baud (see Chapter 8; a 9600-baud link will transmit data at about 1 kbyte/second), IEEE-488 parallel links are often capable of transfer rates up to 1 Mbyte/second. This figure can be exceeded by other special-purpose parallel and serial links.

Regardless of the potential hardware link speed, the actual speed may be substantially lower, depending on your computer's ability to absorb the incoming data and on any additional processing time within the equipment sending the data. To appraise the timing performance of your equipment, it is worth experimenting with a simulation if timing is critical. If very long data acquisition times are involved, you might consider running the experiment overnight.

6.2.3 Compression

You can save both transmission time and storage space after acquisition by compressing your data. Compression techniques and algorithms are well established and widely published, so you can concentrate on implementation. Thankfully, implementing compression need not be difficult nor time-consuming.

Perhaps the simplest form of compression is to eliminate (delete) any data you do not need to keep. For example, if you are storing ASCII numbers, is the number of digits in the number comparable with the accuracy of the data? If not, storing fewer digits might save space. Similarly, a two-byte integer (**Int16** in *Mathematica*, **short int** in C, or **INTEGER*2** in FORTRAN) can take up six bytes (including the sign but excluding any surrounding spaces) when stored as an ASCII number. For high-precision floating-point numbers (say, eight or more bytes in binary format), storage in binary form returns even better value. Thus by formatting your data appropriately, you can save space. You will find examples of formatting data for output in Chapter 5.

How much memory space you can save by using compressed data can be theoretically determined by calculating the entropy of the dataset. Thankfully, entropy is simple both to grasp conceptually and to calculate. Entropy is a measure of the randomness of the dataset – the more random the data are, the more symbols are required to transmit them, and the greater the entropy. For example, the English-language has 26 symbols (*a* to *z*, ignoring case, punctuation, and so on) all of which are used, but some more often than others. If we give each symbol a binary code, then each symbol would need 5 bits to encode it. It would make more sense to encode more frequently used symbols with shorter codes. This principle is used by the Morse code: The symbol *e* (most commonly used) is encoded as a *dot*, whereas the symbols *x* and *j* (less commonly used) are encoded with longer codes (*dash-dot-dot-dash* and *dot-dash-dash-dash* respectively).

Given a set of k symbols, each of which has a probability of use P(k), the entropy E of the symbol set is

$$E = -\sum_{i=1}^{k} P(i) \log_2 P(i)$$

and equals the average number of bits needed to store a member of the dataset. Thus the total number of bits required for storage of a set of k symbols is the product of the entropy value and the number of members in the set. Note that in this context a value is the same as a symbol. A symbol (like the letter a) can occur many times in a set of symbols (like this book), so a value (for example, 8) can occur some number of times in a set of your data.

For example, if your data have 10 possible values (**si** values: 1 through 10) and you count the number of times that each value occurs in a sample of data (**pi**), then you can calculate the entropy of the set.

Here is the information that we need (value 10 occurs 14 times):

In:

```
Si={1,2,3,4,5,6,7,8,9,10};
pi={1,2,6,2,1,7,3,9,1,14};
```

We can then calculate the probability of each value occurring by applying **Plus** to the list **pi** to find the total number of all occurrences, and then by dividing **pi** by the total.

In:

```
pi=pi/Apply[Plus, pi]//N
```

Out:

```
{0.0217391, 0.0434783, 0.130435, 0.0434783, 0.0217391,
 0.152174, 0.0652174, 0.195652, 0.0217391, 0.304348}
```

Next, we define a function to calculate the entropy component of each element in the list of probabilities. By mapping that function over **pi** and summing the resulting list, we obtain the entropy (measured in bits) for our sample.

In:

```
entropyFunction[x_]:=-x Log[2,x];
```

In:

```
Map[entropyFunction, pi]
```

Out:

```
{0.120077, 0.196677, 0.383296, 0.196677, 0.120077,
0.413336, 0.256865, 0.460494, 0.120077, 0.522324}
```

In:

```
entropy=Apply[Plus,%]
```

Out:

```
2.7899
```

Note that the values themselves (the symbols) never appear in the calculation; only the probability of a value occurring is used.

To encode 10 values that have identical probabilities of occurring would take 4 bits – of which only the codes 0000_2 through 0101_2 are needed. With optimal encoding, to encode 10 values with the occurrence probabilities given in **pi** would take only 2.8 bits for the average encoded word length: an instant memory savings of 30%. Using written English text as an example, the entropy is typically just around 4.2 bits, compared with the simple binary-code requirement of the alphabet of 5 bits; the lower value of the entropy implies a space saving of 16% is possible. (Commercial compression programs take advantage of other language features to achieve still higher compression factors.)

The entropy value of the dataset gives us a length for the average word at which to aim; it will not be possible to use fewer bits to store or transmit the data. You may well want to use more bits – a less efficient encoding – if it saves you a lot of effort and time in decoding.

How you encode your data will depend on their nature. If the dataset contains runs of same-valued elements, you can store the run value and the number of elements with that value. For data with long runs, this technique – called *run-length encoding* (RLE) – is quite efficient and very simple to implement either in your *MathLink* code or in *Mathematica*. Run-length encoding in *Mathematica* is discussed by Gaylord et al. (1993), (p. 135), and by Wolfram (1991), (p. 13); the latter gives the following recursive code that makes use of *Mathematica*'s extensive pattern-matching facilities.

In:

```
rld={1,1,1,1,1,1,1,3,4,4,4,4,4,4,4,4,
    4,4,4,4,5,5,5,5,2,2,2,2,2};
RunEncode[{rest___Integer, same:(n_Integer)..}]:=
  Append[RunEncode[{rest}], {n,Length[{same}]}];
RunEncode[{}]:={};
```

In:

```
RunEncode[rld]
```

Out:

```
{{1, 7}, {3, 1}, {4, 11}, {5, 4}, {2, 5}}
```

Run-length encoding is perhaps the simplest of the many encoding schemes that you can implement in *Mathematica* because the result is still a set of integers, stored in a normal manner.

There are many other compression techniques that use a variable number of bits per word. Languages that are well suited to working at the bit-level (like C or FORTRAN) cope with variable-length coded data with a small overhead. *Mathematica* is designed to work with numbers and symbols at a much higher level, and so it is not optimally matched to the task of handling nonstandard-size codes. Fortunately, other compression techniques do not rely on nonstandard code sizes and so are more easily implementable in *Mathematica*, although with some penalty of reduced compression efficiency. For example, suppose you have data that were acquired with a 16-bit analogue-to-digital convertor. A slowly varying signal might have differences between adjacent samples that are within the range -8 to $+7$. Such a small numerical range could be encoded with just 4 bits, so it is wasteful to use 16 bits (which can encode any value in the range -32768 to $+32767$) for every sample. (Of course, we still have to be prepared for sudden jumps in sample value that might well require the full 16 bits.) Although our scheme could achieve a factor of four in space saving, it is simpler for *Mathematica* to deal with bytes rather than half-bytes.

If we compromise on the intersample range so that values between -127 and $+127$ are encoded with 8 bits and we use the value of -128 to signify when the next 16 bits contain an uncompressed value, then we might reduce the volume of data by nearly a factor of two. After any full 16-bit number has been read, we can automatically revert to reading the next 8 bits, looking for either a compressed intersample difference (-127 to $+127$) or another marker (-128). This type of compression code is a modified differential shift code (mDSC). Cappellini (1985) and Gonzalez & Wintz (1977) have published contain useful discussions on a selection of encoding methods. The following explains how we implement mDSC in *Mathematica*.

Our first tasks are to load the standard package that contains the functions to manipulate binary files and to generate some integer-valued test data. We find it useful, too, to plot the data.

In:

```
Needs["Utilities`BinaryFiles`"];
data=Table[Round[3000 Sin[3t/100-1]//N],{t,1,200}];
ListPlot[data]
```

Out:

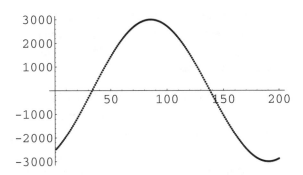

We can see what the differential values are like by subtracting from any value its predecessor value: $\delta_i = d_i - d_{i-1}$. To achieve this operation, we can cheat slightly and simply rotate the whole of the data list to the right before carrying out the subtraction. For all but one of the data members, rotation produces the correct result; the last member of the list, however, is subtracted from the first member – a meaningless difference. On the other hand, we want only a quick look, and the first difference is undefined in any case (we do not know what value preceded the first of our samples). Note that the plot of the differences has a large first value. (If you wanted to remove the meaningless difference, you could use the function **Drop[data-RotateRight[data], 1]** to delete the list's first value.)

In:

```
ListPlot[data-RotateRight[data]]
```

Out:

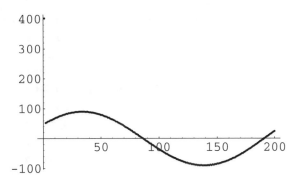

Now that we have some data, we can save them in a binary file using 16-bit integer format. The raw, uncompressed dataset has 400 bytes.

In:

```
outStream=OpenWriteBinary["raw data"];
WriteBinary[outStream, data,
    ByteConversion ->(ToBytes[#,
                      IntegerConvert -> Int16]&)];
Close[outStream];
```

To compress the data, we write a function **storeCompressed** that takes two arguments: the dataset and the name of a file into which the data will be stored. The function workings are kept simple. First, the output file is opened. Then, for each member of the dataset, we write out either the difference between it and the preceding value (if that difference is within the range -127 to $+127$), or else a flag (-128) and the full 16-bit value. Lastly, we close the file.

We then use **storeCompressed** to store **data**.

In:

```
Clear[storeCompressed];
storeCompressed[data_List,file_String]:=
  Module[{lastValue=99999999,
          flagValue=-128,i,outStream},
        outStream=OpenWriteBinary[file];
        For[i=1, i<=Length[data],i++,
            If[Abs[data[[i]]-lastValue]<=127,

                WriteBinary[outStream,
                            data[[i]]-lastValue,
                            ByteConversion->
                              (ToBytes[#,
                               IntegerConvert->Byte]&)];,

                WriteBinary[outStream,
                            flagValue,
                            ByteConversion->
                              (ToBytes[#,Byte]&)];
                WriteBinary[outStream,
                            data[[i]],
                            ByteConversion->
                              (ToBytes[#,
                               IntegerConvert->Int16]&)];
                ];
            lastValue=data[[i]];
          ];
        Close[outStream];
        ];
```

In:

```
storeCompressed[data,"compressed data"]
```

The compressed data take up 202 bytes in the file. For this dataset, we have nearly halved the storage required.

To retrieve our data, we need to write an expansion function, **readCompressed**. The task of decompressing is also simple. We open the file and read a byte. If that byte contains the flag value (−128), then we read the next two bytes as a 16-bit integer, otherwise we add that byte's value to the preceding value. We use a **While** loop, collecting data for as long as the value the program retrieves from the file is a number; the end-of-file marker terminates the data retrieval. Lastly, we close the file and return the list of values.

In:

```
Clear[readCompressed];
readCompressed[file_String]:=
  Module[{inValue,flagValue=-128,inStream,data={},
          lastValue},
        inStream=OpenReadBinary[file];
        inValue=ReadBinary[inStream,
                           SignedByte];
        While[NumberQ[inValue],
            If[inValue==flagValue,

                inValue=ReadBinary[inStream,
                                   SignedInt16];

                AppendTo[data,inValue];
                lastValue=inValue;,

                newValue=inValue+lastValue;
                AppendTo[data,
                        newValue];
                lastValue=newValue;

            ];

            inValue=ReadBinary[inStream,
                               SignedInt8];
        ];

        Close[inStream];
        data
        ];
```

For any compression/expansion function, we need to check that the recovered data are indeed the same as the original dataset. In addition to plotting the recovered data, we can calculate the difference between the original data and the recovered data and verify that it is zero-valued.

In:

```
readInData=readCompressed["compressed data"];
ListPlot[readInData]
```

Out:

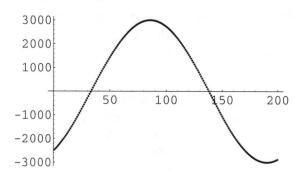

By using the output modifier **Short**, we can get a summary of the values resulting from subtracting **data** from **readInData**; the result is 10 zeros, followed by 185 other terms, and terminated with 5 zeros.

In:

```
readInData-data//Short
```

Out:

```
{0, 0, 0, 0, 0, 0, 0, 0, 0, 0, <<185>>, 0, 0, 0, 0, 0}
```

In:

```
Apply[Plus, Abs[readInData-data]]
```

Out:

```
0
```

An extension to the modified differential shift coding scheme is to include a predictor stage. The predictor is typically a simple extrapolator that uses, say, the last two or three values to predict the next value. The encoder then works on the difference between the current and predicted values rather than between the current and previous values.

Predictor-based compressors cope well with slowly varying data – where the δ_i would otherwise exceed the compressed range (±127) – but cope poorly with very noisy or truly random data. Encoding schemes that rely on *a posteriori* statistical analysis of the data – Huffman codes, for example (discussed by Gaylord et al. (1993)) – are more efficient for noisy or random data because they do not rely upon or use any assumptions about the position of values within the dataset.

Here is an example of how to generate a Huffman code tree. We first define a list of symbols and their probabilities. Each element in the list has three components: *i*) the probability of that symbol being used, *ii*) the symbol itself, and *iii*) a blank list in which that symbol's code is built up.

In:

```
pri={{0.5,a,{}},{0.2,b,{}},{0.15,c,{}},{0.1,d,{}},
    {0.05,e,{}}};
```

We can check that the sum of the symbol probabilities is indeed 1 just by applying the **Plus** function to **pri**. A by-product is a sum of all the symbols, too.

In:

```
Apply[Plus,pri]
```
Out:

```
{1., a + b + c + d + e, {}}
```

The generation of the Huffman code tree is relatively straightforward. The list of symbols and the probabilities of their occurrence is sorted into ascending order. The two smallest probabilities are then summed, removed from the list, and replaced, in a combined entry, into the (revised) list. As long as the revised list has more than one member, this process is repeated, and building-up the tree with the codes **c0** and **c1**.

In:

```
Clear[makeHuffman];
makeHuffman[pri_]:=Module[{spri,sum},
  spri=Sort[pri];
  While[Length[spri]>1,
        AppendTo[spri[[1,3]],c1];
        AppendTo[spri[[2,3]],c0];
        sum={spri[[1,1]]+spri[[2,1]],
            {spri[[1,2]],spri[[2,2]]},
            {spri[[1,3]],spri[[2,3]]}};
        spri=Drop[spri,2];
        PrependTo[spri,sum];
        spri=Sort[spri];
        ];
        spri
    ];
```

Now we can see the resulting codes for **pri**.

In:

```
makeHuffman[pri]
```

Out:

```
{{1.,
{{b, {c, {e, d}}}, a},
{{{c1}, {{c1}, {{c1}, {c0}, c0}, c0}, c1}, {c0}}}}
```

Parts 2 and 3 of the result give tree-formatted tables of the symbols and their Huffman codes, respectively. For example, symbol **a** has the code **c0** (the codes **c0** and **c1** might correspond to a single bit, set with values 0 and 1, respectively), **b** is **c1c1**, **c** is **c1c0c1**, and so on.

Note that differential or predictor-differential encoding (with fixed or variable-length codes) retains data integrity. You can recover all your data values without error. If your application can tolerate loss of integrity (for example, if small changes in value can be ignored), then you might consider a modified run-length encoder that treats samples with differences smaller than some limit compared with the sample that initiated that run as *bona fide* members of the run.

If you are working with thresholded data (that is, data whose values are below or above some value and thus are of no interest), you might want to run-length encode the data, modifying the usual scheme to redefine a run as one or more samples that meet the thresholded criterion and where samples in the run need not have the same value (the normal criterion for run membership in classical run-length encoding schemes). In datasets that have many identically valued members, classical run-length encoding is very efficient; the modified encoding is better suited when data thresholding is in operation and acceptable data are not necessarily often identically valued.

Here is a non recursive, procedural section of *Mathematica* code that gives a modified run-length encode of the data in **runs**. For a run of data values that are greater than the given threshold, the code lists the starting position, then the data values, then a **-1** to flag the end of the run.

In:

```
runs={1,1,1,2,3,4,5,6,5,4,3,2,1,1,1,
      1,1,1,1,2,3,4,6,7,8,7,6,5,4,3,2,1,0};
mRLE={};
inRun=False;
threshold=2;
placeInDataSet=0;
Map[Function[placeInDataSet++;
             If[inRun,
                If[#>threshold,
```

```
                    AppendTo[thisRunList,#],

                    AppendTo[thisRunList,-1];
                    AppendTo[thisRunRecord,
                            thisRunList];
                    AppendTo[mRLE,
                            thisRunRecord];
                    inRun=False;
                    ],
                If[#>threshold,

                    inRun=True;
                    thisRunList={#};
                    thisRunRecord={placeInDataSet};
                    ]
                ]
            ]
        ,runs];
    Flatten[mRLE]
```

Out:

```
{5, 3, 4, 5, 6, 5, 4, 3, -1, 21, 3, 4, 6, 7, 8, 7, 6,
  5, 4, 3, -1}
```

Other compression techniques that use variable-length words (such as Fano and Huffman codes) might offer you greater space savings, but the compressed data may not be as simple to decompress and can be susceptible to errors in a way that may make it difficult for you to recover the original data. Sophisticated compression techniques are best used in conjunction with error correcting codes (see Castellini (1985), Gonzalez & Wintz (1977), Hill (1986), and Press et al. (1994)).

6.2.4 Binary and ASCII forms

As we have shown in Chapters 1 and 5, *Mathematica* can read or write data in either ASCII or binary format. Which format you use will depend upon a number of factors:

- ASCII files can be read by eye and, where other computer programs have to read or write the data, may offer the simplest method for data interchange.
- Binary files are more efficient for storing floating-point and large integer numbers.
- Interchanging data between programs using a binary format requires the receiving program to know the binary format (for example, 4-, 8-, 12-byte length for floating-point numbers) in which the data have been stored.

- *Mathematica*'s **Utilities`BinaryFiles`** package copes with the most common formats for binary data. By using a simple *MathLink* program, however, you can read binary data in whatever format the compiler on your computer supports. (Before you can transmit a nonstandard format number to *Mathematica* your program will need to convert the number to one of the standard formats. In C, such conversions are implemented using casting. You will need to consult your C compiler's documentation to find out which floating-point types correspond to the 4- and 8-byte IEEE standard formats. For example, the Symantec C/C++ compiler for the Macintosh defines the type **short double** to be the same as the IEEE 8-byte floating-point number format. (By default, the base type **double** normally uses the native floating-point format of any hardware numeric co-processor that is present; on a Macintosh, **double** corresponds to a 12-byte floating-point format, for example, but on a Power Macintosh only the 8-byte IEEE standard format is available in native mode with the Symantec native compiler.)

Here is a small section of C code for the Symantec C compiler for the Macintosh that would convert from 12-byte to 8-byte floating-point formats:

```
/* declare variables */

double myFormatFloat;
short double myStandardFloat;
.
.
.
/* apply cast to change type */

standard=(short double)myFormatFloat;
.
.
.
```

6.3 Safety in laboratory systems

The principal concern for any system designer or builder should be the safety of the people who will use the system, regardless of how frequently they use it. In the rush to publish results, or to meet funding-authority or client deadlines, it can be easy for safety to become a low priority. Laboratory environments can be dangerous enough without further hazards being introduced by computerized machines.

Thankfully, you can enhance safety with a small amount of effort early in the design phase. For example, if your experiment uses a high-voltage supply, you should ensure that the users have to carry out some deliberate act to apply power to the supply. Also, in the event of a power cut, you should ensure that the high-voltage supply remains off until your user deliberately switches it on.

You also should consider the effect of a computer crash on potentially dangerous equipment. Could the computer send a sporadic control code that would set the high-voltage supply to its maximum limits or change its polarity?

When you design the user interface, it is also a good idea to always provide the user a way out of – an escape from – any situation. For example, on most graphical user interfaces, dialogs have a cancel option. Remember to make sure that any cancellation leaves the system in a clean and safe state.

Are there any values that when entered by the user would cause a problem? If so, always check for valid entries. Similarly, you must always confirm that computer-generated output (a voltage or a force on a hydraulic ram, say) is correct before you connect any samples or materials.

6.4   References

Booch, G., "Object-Oriented Analysis and Design (second edition)," Benjamin/Cummings Publishing Company Inc., Redwood City, California, USA, 1994.

Cappelini, V., editor, "Data Compression and Error Control Techniques with Applications," Academic Press, London, United Kingdom, 1985.

C Users' Group, "The C Users' Group Library (CD-ROM)," Walnut Creek CD-ROM, Walnut Creek, California, USA, 1994.

Gaylord, R. J., Kamin, S. N., Wellin, P. R., "Introduction to Programming with *Mathematica*," TELOS Springer-Verlag, Santa Clara, California, USA, 1993.

Gonzalez, R. C., Wintz, P., "Digital Image Processing," Addison-Wesley, Reading, Massachusetts, USA, 1977.

Hill, R., "A First Course in Coding Theory," Oxford University Press, Oxford, United Kingdom, 1986.

Horowitz, E., Sahni, S., "Fundamentals of Data Structures," Pitman, London, United Kingdom, 1982.

Press, W. H., Flannery, B. P., Teukolsky, S. A., Vetterling, W. T., "Numerical Recipes (second edition)," Cambridge University Press, Cambridge, United Kingdom, 1994.

Sexton, C., "Newnes C++ Pocket Book," Butterworth-Heinemann Ltd., Oxford, UK, 1993.

Shlaer, S., Mellor, S. J., "Object-Oriented Systems Analysis: Modeling the World in Data," Prentice-Hall Inc., Englewood Cliffs, New Jersey, USA, 1988.

Wolfram, S., "*Mathematica* — A System for Doing *Mathematics* by Computer," Addison-Wesley, Redwood City, California, USA, 1991.

CHAPTER 7

Understanding *MathLink*

MathLink is the protocol used to send *Mathematica* expressions between the "notebook front end" (where users enter expressions and output is displayed) and the kernel (where the expressions are evaluated). *MathLink* can transport *Mathematica* expressions between programs running on the same or different computers; it even works when the computer network is heterogeneous! The *MathLink* protocol is part of *Mathematica* and is also deployed as a portable C language library that you can use in your own programs. Why would you want to use *MathLink*?

- You have an algorithm that needs to be implemented in a compiled language for efficiency reasons (even though *Mathematica* is extremely efficient for many numerical computations).
- You need to use a preexisting library of functions and you do not have source code. For example, you are using specialized laboratory hardware such as a multifunction A/D-D/A-DIO board that is controlled through a library supplied by its manufacturer.
- You have "legacy" libraries that you do not want to rewrite in *Mathematica*.
- You want to call the *Mathematica* kernel as a computational engine from within another program.

The first three examples mentioned above are cases where you want *Mathematica* to call your code, that is, where you want to extend *Mathematica* with additional functions. Extending *Mathematica* is quite easy because there are tools to assist you in building such plug-ins. The last example mentioned above is the case where you want to call *Mathematica* from within a "front end" you have written. There are no tools provided with *Mathematica* to assist you in this case. Unless you have a preexisting program that you wish to extend, or if you are experienced in the design of user interfaces, you probably will not do much calling of *Mathematica* from your own program.

The primary documentation for *MathLink* is found in *The* Mathematica *Book*, third edition. You can also find the documentation in the online `Help Browser` in *Mathematica* 3.0.

Now, because the shipping version of *Mathematica* is 3.0 or later, this chapter refers to Version 3.0 of *MathLink*. However, most of the information is also correct for earlier versions, although a few of the functions and features may not be present. The programs presented here should also work with future versions of *MathLink*; the *MathLink* development team at Wolfram Research strives to avoid ever "breaking" a correctly written *MathLink* program.

It is one thing to write or use an external function, but how do you teach *Mathematica* when and how to use your functions? You use *Mathematica*'s **Install** function to incorporate external functions into the *Mathematica* environment. Such external functions are termed "installable".

Installable functions are actually separate programs that communicate with *Mathematica*. They usually run on the same machine as *Mathematica*, but they can run just as easily on another networked machine of the same or different type. *Mathematica* can maintain more than one open link, communicating with more than one external program. For example, if you have several IEEE-488 controllers (computers with an IEEE-488 interface board) scattered across your laboratory, you could design a centralized control and analysis system using *MathLink*'s communication capability.

How do you make installable functions? Ideally you should be able to create installable functions from preexisting functions with minimal effort. You probably do not want to modify your functions and do not want to write new code. It has long been common practice among experimenters to stitch together functions from preexisting libraries (with a minimal amount of programming). Such applications usually have what would charitably be called "rustic" user-interfaces (often of the print-a-line/input-a-number teletype variety that would not have been out of place in the '60s). Hopefully, the previous chapters of this book have convinced you that *Mathematica* is an excellent tool both for performing analyses (with rich graphics) and for writing a fully annotated report or paper. This chapter will show that *MathLink* allows you to have the rich tool set and interface of *Mathematica* with the relative ease of the older cut-and-glue approach.

Automated assistance in making functions installable is desirable since the compiled languages used for most libraries are more difficult to master than *Mathematica*. This is also essential when you are using supplied libraries without source code. Is it a fantasy like self-diapering infants? No! The extra code required to call external functions from *Mathematica* is so stylized and abstractable that the *MathLink* team at Wolfram Research was able to design a simple declarative template language and associated template preprocessor to automatically generate most (and often all) of the wrappers

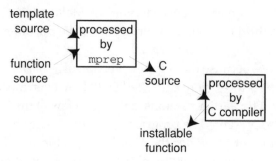

Figure 7a   Production of installable functions

and scaffolding necessary. Templates are platform-independent; your extensions to *Mathematica* will therefore be as portable as the external code that you write or borrow from a library.

A template specifies the name of the function, the arguments that *Mathematica* should pass to the function, the types of the arguments, and the type of the function's return value. The template file is then preprocessed by a tool called mprep, which generates C code that manages most (possibly all) of the *MathLink*-related aspects of the program. Templates are platform-independent; you simply choose the version of mprep for the platform on which you wish to run your external function.

The big picture is illustrated in Fig. 7a

You can group one or more installable functions together into a single installable program. When you **Install** the program, you make all of its functions available to *Mathematica* at the same time.

7.1   The pons asinorum: addtwo

We look first at an extremely simple example of an installable program, the **addtwo** program supplied with the *MathLink* Developer's Kit. The goal of this example is to write an installable function that adds two integers and returns the sum. Here is the C source file addtwo.c:

```
#include "mathlink.h"
    int addtwo(int i, int j) {
        return i+j;
    }
    int main(int argc, char* argv[]) {
        return MLMain(argc, argv);
    }
```

The function **addtwo** is obvious; it is a C routine that takes two ints and returns the sum. main is a function that every C program must contain; it is the entry point at which program execution begins. In this case, we have written a "stub" that simply calls the real main function (named MLMain), which is automatically generated by mprep. (Microsoft Windows users will see a main function that is slightly more complicated.)

mprep generates MLMain() instead of main() because you might need an opportunity both to "set up" before MLMain() and to "clean up" after MLMain().

Here is the corresponding template file addtwo.tm:

```
:Begin:
:Function:        addtwo
:Pattern:         AddTwo[i_Integer, j_Integer]
:Arguments:       { i, j }
:ArgumentTypes:   { Integer, Integer }
:ReturnType:      Integer
:End:
```

What does this mean? The :Function: line specifies the name of the C routine you want to call from *Mathematica*. The :Pattern: line shows when the routine will be called. The pattern given on this line will be translated into the left-hand side of a function definition, exactly as you would type it if you were creating the entire function in *Mathematica*. The :Arguments: line specifies the expressions to be passed to the external program. These expressions do not have to be the same as the variable names on the :Pattern: line, although they often will be. You could, for example, put **{Abs[i], j^3}**. The point is that what you put on the :Pattern: line and the :Arguments: line is *Mathematica* code; it will be used verbatim in a definition that could be caricatured as follows:

```
AddTwo[i_Integer, j_Integer] :=
    SendToExternalProgramAndWaitForAnswer[{i, j}]
```

The :ArgumentTypes: and :ReturnType: lines contain special keywords used by mprep to create the appropriate *MathLink* calls that transfer data across the link.

The details of building the executable from the addtwo.tm and addtwo.c source files differ from platform to platform (and from compiler to compiler on a given platform). On UNIX, you will usually use the mcc script that comes with *Mathematica*. In this example the appropriate command for mcc is:

```
mcc addtwo.tm addtwo.c -o addtwo
```

The two steps that `mcc` performs are as follows: *i*) run `mprep` on the `.tm` file to create a `.tm.c` file, and *ii*) compile and link all the source files, including the `.tm.c` file, specifying to the `cc` compiler where to find the `mathlink.h` file and the *MathLink* library file (named `libML.a` on UNIX machines). It is the `.tm.c` file that contains the `mprep`-generated C source. Normally, this file is deleted by `mcc` after it has been compiled; if you want to see what it looks like, you can prevent its deletion by specifying the `-g` command-line option to `mcc`. Advanced users of *MathLink* can learn a lot by studying this file. On MacOS and Windows systems, the steps to build the program will be different, and you should consult the README file that comes with your *MathLink* Developer's Kit.

The `mcc` method is convenient for simple projects, but it has some drawbacks, one of which is that it is hard coded to call the `cc` compiler. You might want to skip `mcc` altogether and write your own makefile. In that case, you will be calling `mprep` yourself. Here is an example:

```
/math/Bin/MathLink/mprep addtwo.tm -o addtwo.tm.c
```

Note that `mprep` is not on your UNIX path, so you will need to specify the full pathname. The *MathLink* library, `libML.a`, is also located in the `math/Bin/MathLink` directory, and the `mathlink.h` file is in the directory `math/Source/Includes`.

To use the **AddTwo** function in *Mathematica*, you launch the external program with *Mathematica*'s **Install** function, which takes the name of the program generated by `mcc` or your Windows or Macintosh code development system.

In:
```
link = Install["addtwo"]
```
Out:
```
LinkObject[addtwo, 2, 2]
```

The function **LinkPatterns** shows what functions are defined by the external program associated with a given link.

In:
```
LinkPatterns[link]
```
Out:
```
{AddTwo[i_Integer, j_Integer]}
```
In:
```
AddTwo[3,4]
```
Out:
```
7
```

You may wonder how the definition for **AddTwo** appears in *Mathematica*. After all, all we have done is start up the kernel and type **Install** yet suddenly *Mathematica* knows about a function called **AddTwo**. The answer is that the external program sends to *Mathematica* the definitions for the functions it exports when the link is first opened. Here is what such a definition looks like:

In:

```
??AddTwo
```

Out:

```
AddTwo[i_Integer, j_Integer] :=
   ExternalCall[LinkObject["addtwo", 2, 2],
                CallPacket[0, {i, j}]]
```

Of course, the programmer never sees any of this because it is handled at one end by the code that mprep writes and at the other end by the **Install** function. Most programmers have no reason to care how this feat is performed, but you should know that all the code involved is accessible. If you are interested, you might want to take a look at a .tm.c file. You can see how the built-in **Install** function is implemented by typing **??Install** from within *Mathematica*.

7.2   Using :Evaluate: **to include *Mathematica* code**

When the external program is installed it sends code to teach *Mathematica* how to call the functions defined in the templates. Your template also can include arbitrary *Mathematica* code to be sent. For example, you might have some accessory code – perhaps a usage message – that your functions rely upon and need defined in *Mathematica*.

You can specify arbitrary *Mathematica* code to be sent to the kernel when your program is installed by using another feature of template files, the :Evaluate: line. Here is an example of specifying a usage message:

```
:Evaluate:       AddTwo::usage = "AddTwo[i, j] adds two
integers."

:Begin:
:Function:       addtwo
:Pattern:        AddTwo[i_Integer, j_Integer]
:Arguments:      { i, j }
:ArgumentTypes:  { Integer, Integer }
:ReturnType:     Integer
:End:
```

Defining messages is a trivial use of :Evaluate: lines. Another common use is to make your functions appear in a package context. The current behavior of **Install** is to cause all functions defined in installable programs to appear in the **Global`** context, not the current *Mathematica* context. (Note that this behavior may be changed in a future version.) This means that if you want the **AddTwo** function to appear in a package context, such as **MyPackage`**, then you cannot do this:

In:

```
BeginPackage["MyPackage`"];
Install["addtwo"]
EndPackage[]
```

in this example, the **AddTwo** function will still be put into the **Global`** context. The best way to handle this is to put the **BeginPackage** statement into an :Evaluate: line in the .tm file:

```
:Evaluate:        BeginPackage["MyPackage`"]
:Evaluate:        AddTwo::usage = "AddTwo[i, j] adds two
integers."
:Evaluate:        Begin["Private`"]

:Begin:
:Function:        addtwo
:Pattern:         AddTwo[i_Integer, j_Integer]
:Arguments:       { i, j }
:ArgumentTypes:   { Integer, Integer }
:ReturnType:      Integer
:End:

:Evaluate:        End[ ]
:Evaluate:        EndPackage[ ]
```

Everything that follows an :Evaluate: up until the first blank line or line whose first character is not a space will be sent as a single unit. This means you need to have a separate :Evaluate: for each separate statement or definition. We discuss further the use of :Evaluate: later in this chapter.

7.3   Putting and getting arguments manually

Note that no *MathLink* calls were needed in writing the addtwo program and its template. With a little additional effort, you can take more control over the passing of arguments and return values. This would be necessary, for example, if the installable function needed to receive or return expression types that are not among the set handled automatically by mprep, or if the function returned different types of results (such as an integer or the symbol **$Failed**) in different situations.

As an example, we will modify the addtwo program so that it works for larger integers, up to the long integer size. This is a useful task since some computer – compiler combinations limit variables of type int to the range −32768. . .32767. In the template file, the keyword Integer on the :ArgumentTypes: and :ReturnType: lines causes mprep to create calls to *MathLink* routines **MLGetInteger** and **MLPutInteger**, which transfer C ints. Instead, we need to call **MLGetLongInteger** and **MLPutLongInteger**, so we change these two lines:

```
:ArgumentTypes:   { Manual }
:ReturnType:      Manual
```

The keyword Manual on the :ArgumentTypes: line informs mprep that we are responsible for writing our own calls to get the arguments; similarly, Manual on the :ReturnType: line indicates that we will put the result ourselves. Here is how the addtwo function looks now:

```
void addtwo(void)
{
  long i, j, sum;

  MLGetLongInteger(stdlink, &i);
  MLGetLongInteger(stdlink, &j);
  sum = i + j;
  MLPutLongInteger(stdlink, sum);
}
```

Note the change in the function's prototype. For those of you who are not familiar with C, the function declaration void addtwo(void) means that the function takes no arguments and does not return a result. Remember that the actual call to the addtwo function is made from code that mprep writes, so the function's arguments and return value must match mprep's assumptions, as determined from the :ArgumentTypes: and :ReturnType: lines of the template. By specifying Manual on the :ArgumentTypes: line, you tell mprep to pass no arguments to addtwo when it is called. Similarly, by specifying Manual on the :ReturnType: line, you tell mprep to ignore any return value. In other words, in our first example addtwo communicated via the C runtime stack to a hidden layer of code generated automatically by mprep; the hidden layer was then responsible for handling all *MathLink* calls. In this second example, we must handle the communication via explicit *MathLink* calls.

Note that we are still using a large amount of hidden functionality from the mprep-generated code. The variable stdlink is automatically generated and initialized by mprep. C programmers will notice that stdlink is analogous to the file descriptors stdin and stdout. In each case, you are provided with these channels without having to explicitly ask for them. Similarly, MLPutx and MLGetx calls are analogous to printf and scanf because you use them to Put/print and Get/scan, respectively. You will find that your instincts and experience using those functions will serve you well with simple *MathLink* programming (up to a point!).

It is possible to use Manual on one of these lines and not the other. It is also possible to mix Manual with other types on the :ArgumentTypes: line. For example, if you want to automatically read the first argument but get the second one yourself, you can write:

```
:ArgumentTypes:   { Integer, Manual }
```

In this case, the addtwo function would be written to take one int argument, and inside it there would be one call to **MLGetInteger**. If you use Manual on the :ArgumentTypes: line, it must be the last type in the list. In effect, Manual means "I want to get all the remaining arguments from the link myself." You cannot specify, for example,

```
:ArgumentTypes:   { Integer, Manual, Integer }
```

Because a typical laboratory application will require you to send or receive lists of Integers or Reals rather than a strictly enumerated set, we will now consider how to handle this important case.

7.4   Passing lists and arrays

mprep supports :ArgumentTypes: of IntegerList and RealList, so sending these to your installable function is almost as simple as the case of addtwo. However, :ReturnType: cannot be either of these since both lists expand to two function parameters and C functions can return only a single parameter. Therefore you need to use the Manual keyword (and make explicit calls to the *MathLink* library) when you need to return a list to *Mathematica*. The *MathLink* sample program bitops demonstrates this. For our purposes, only one of the functions defined in bitops.c is relevant: the function complements, which takes a list of integers and returns a list of the bitwise complements of the integers. Here is the template entry in `bitops.tm`:

```
:Begin:
:Function:        complements
:Pattern:         BitComplements[x_List]
:Arguments:       {x}
:ArgumentTypes:   {IntegerList}
:ReturnType:      Manual
:End:
```

There is a keyword `IntegerList` that can be used on the `:Argument-Types:` line; `mprep` will automatically generate C code to get the list for you. You cannot use `IntegerList` in the `:ReturnType:` line; you have to use `Manual` and put the result list onto the link yourself. Here is the C function implementing `complements`:

```
void complements(int px[ ], long nx)
{
   long     i;
   int      *cpx;

   cpx = (int *) malloc(nx);
   for(i = 0; i < nx; i++) cpx[i] = ~ px[i] ;
   MLPutIntegerList(stdlink, cpx, nx);
   free(cpx);
}
```

Note that we have specified only one argument, an `IntegerList`, to be passed to the external function, but that the function itself is written to take an integer array followed by a long integer. Confusion over this is a source of many user errors. When the `mprep`-generated code reads the list of integers, it will determine the length of the list and pass this to your function. Sometimes users mistakenly believe that they must themselves pass the length of the list from *Mathematica*, so they erroneously write the `:Arguments:` and `:ArgumentTypes:` lines like this:

```
:Arguments: {x, Length[x]}
:ArgumentTypes: {IntegerList, Integer}
```

In the arguments to your function, the long parameter that will receive the length of the list always comes immediately after the list itself . For example, if you need to receive a list of integers, a list of reals, and an integer, you would write the `:ArgumentTypes:` line like this:

```
:ArgumentTypes: {IntegerList, RealList, Integer}
```

and the function prototype would look like

```
void myfunc(int ilist[ ], long ilen, double rlist[ ],
            long rlen, int j);
```

To put the result list back to *Mathematica*, you can use **MLPutIntegerList** or **MLPutRealList**.

In addition to putting and getting lists of integers and doubles, *MathLink* has functions for putting and getting multidimensional arrays in a single step, for example, **MLGetDoubleArray** and **MLPutDoubleArray**. You will be interested in these functions if your scientific application involves a data stream or file format that has implied dimensionality. For example, a data stream from a scanning electron microscope might be 200,000 bytes per scan with the data accumulated by scanning over a 400-by-500 field. You are likely to want to send *Mathematica* a two-dimensional list suitable for the **ListDensityPlot** or **ListPlot3D** functions. Check the mathlink.h header file for the complete set of array handling functions. The easiest way to describe the general behavior of these functions is for us to show a sample program. The following is an example function that creates an identity matrix of size n:

```
void identity_matrix(int n)
{
    long    dimensions[2];
    char    *heads[2] = {"List", "List"};
    long    depth = 2;
    int     *mat;
    int     i,j;

    mat = (int*) malloc(n * n * sizeof(int));

    for(i=0; i<n; i++)
        for(j=0; j<n; j++)
            if(i == j)
                mat[i + j * n] = 1;
            else
                mat[i + j * n] = 0;

            dimensions[0] = dimensions[1] = n;

            MLPutIntegerArray(stdlink, mat, dimensions,
                              heads, depth);

            free(mat);
}
```

The `Array` functions are similar to their `List` counterparts. In a `PutArray` function, instead of a `long` length parameter, you pass an array of long integers giving the length in each dimension. The `heads` parameter is an array of `char*` that give the heads in each dimension (**List** in most cases). If the `heads` are **List** in each dimension, you can simply pass `NULL` in place of `heads`. When would any of the heads not be **List**? When you want to pass a list or array of **Complex** or **Rational** or any other *Mathematica* type not found in C but expressible as a C structure.

Here is a complete example showing the use of `MLGetDoubleArray` and `MLPutDoubleArray`. The function transposes a matrix of reals:

```
:Begin:
:Function:              transpose
:Pattern:               MyTranspose[1_?MatrixQ]
:Arguments:             {1}
:ArgumentTypes:         {Manual}
:ReturnType:            Manual
:End:

void transpose(void)
{
   long    *dimensions;
   char    **heads;
   long    depth;
   double  *data;
   int     i, j;
   double  *tdata;     /* put the transposed array here */
   long    tdimensions[2];  /* reverse of dimensions */

   MLGetDoubleArray(stdlink, &data, &dimensions, &heads,
                   &depth);

   tdata = (double*)
   malloc(sizeof(double)*dimensions[0]*dimensions[1]);

   for(i=0; i<dimensions[0]; i++)
      for(j=0; j<dimensions[1]; j++)
         tdata[i + j * dimensions[0]] =
                       data[j + i * dimensions[1]];

         tdimensions[0] = dimensions[1];
         tdimensions[1] = dimensions[0];

         MLPutDoubleArray(stdlink, tdata, tdimensions,
                       heads, 2);
```

```
                    free(tdata);
                    MLDisownDoubleArray(stdlink, data, dimensions,
                                        heads, depth);
        }
```

Note the call to `MLDisownDoubleArray`. Whenever you use `MLGet` to receive an object whose size cannot be known at compile time (for example, a string, a symbol, a list, or an array), *MathLink* allocates memory on its own, reads the data into this memory space, and gives you only the address of the data. For example, in `MLGetString`, you pass the address of a `char` pointer (that is, a `char **`), and *MathLink* stuffs the address of the string it received into your `char*`. You do not have to allocate any memory yourself – or to worry about the data size. At this point, *MathLink* owns the data, and it is waiting for your permission to free the memory that it occupies, which you grant when you call the `MLDisown` functions. Between the time you call `MLGet` and `MLDisown`, you can only read the data. Do not try to modify it *in situ*. If you need to modify the data *in situ*, allocate your own memory and copy the data into it (for example, using the standard C library function `strcpy`).

(Note to advanced programmers: If you need to use your own custom memory allocation functions, you can teach *MathLink* to use them instead of its own. Theo Gray, designer of the *Mathematica* notebook interface, used this technique to ensure that the notebook front-end's memory allocators were used by *MathLink*.)

Note that you need to worry about calling `MLDisown` functions only if you call `MLGet` yourself. For strings, symbols, and lists that `mprep` gets for you automatically, it takes care of calling the appropriate `MLDisown` functions after your function returns.

7.5   Passing arbitrary expressions

MathLink has functions for passing all native C types, along with single and multidimensional arrays. There are times, though, when you need to send or receive expressions that do not fit neatly into C types. Your function might need to return a list of mixed integers and reals, or a list of lists that is not a matrix, or a symbolic expression like **Integrate[x^2, {x,0,1}]**. How do you go about transferring expressions like these?

Here, we focus on returning such expressions from an external function. It is less likely that your function would want to receive such expressions. It is certainly possible to receive complex expressions, but what would you do with them? You would have to write your own code to analyze them and then extract the desired information. If you need to deal with complicated expressions in your external functions, you would be better off writing some code on the *Mathematica* side that acts as a wrapper around your template functions, manipulating and decomposing the expressions into meaningful C-size chunks, and sending these instead. This type of chore is more easily programmed in the *Mathematica* language than in C.

You send expressions over *MathLink* in a way that mimics their **Full-Form** representation. There are *MathLink* functions for the necessary atomic types (integers, reals, strings, and symbols), and if you need to put a composite expression (something with a head and zero or more arguments), you use MLPutFunction to put the head and the number of arguments, then MLPut calls for each of the arguments in turn. For example, to put **Integrate[x^2, {x,0,1}]** onto a link, you would write:

```
MLPutFunction(stdlink, "Integrate", 2);
  MLPutFunction(stdlink, "Power", 2);
      MLPutSymbol(stdlink, "x");
      MLPutInteger(stdlink, 2);
  MLPutFunction(stdlink, "List", 3);
      MLPutSymbol(stdlink, "x");
      MLPutInteger(stdlink, 0);
      MLPutInteger(stdlink, 1);
```

Of course, if you want to return an expression like this from your function, you will need to declare a Manual return type in the .tm file.

A very common error is attempting to put more than one expression from the external function. An external function, like any built-in function, cannot return two things. In the earlier examples, we sent complex expressions back to *Mathematica*, but there was always only one of them. Here is an example of the error of sending more than one expression:

```
void return_two(void)
{
  int i, j;

  MLGetInteger(stdlink, &i);
  MLGetInteger(stdlink, &j);

  MLPutInteger(stdlink, i);
  MLPutInteger(stdlink, j); /* oops, returned an extra
                                value */
}
```

The two integers returned would need to be wrapped in a head of some sort so that they become a single compound expression. If you had intended to return the list **{i,j}**, write the put calls like this:

```
MLPutFunction(stdlink, "List", 2);
    MLPutInteger(stdlink, i);
    MLPutInteger(stdlink, j);
```

7.6   Requesting evaluations by the kernel

The external function can request evaluations by *Mathematica* between the time it is called and the time it returns its result. For example, you might want *Mathematica* to assist you in computing something, or you might want to trigger some side-effect such as displaying an error message. The *MathLink* function MLEvaluate is designed for this purpose. MLEvaluate takes a string argument that will be interpreted by *Mathematica* as input. The result will be returned to your function as an expression wrapped with the head ReturnPacket. You should read this ReturnPacket from the link whether you care what is in it or not.

Use such callbacks judiciously; if you call *Mathematica* very often from your external compiled function, you will find that the function runs no faster than *Mathematica* itself! Nonetheless, such callbacks are invaluable in integrating external code into *Mathematica*. The installable *MathLink* program MathHDF (Janhunen & Stein (1993)) uses MLEvaluate to map numeric recoverable error codes returned by the HDF library into recognizable *Mathematica* messages.

As a simpler example, we go back to the addtwo function and modify it to detect an overflow when adding the two long integers (that is, a sum that is outside the range of a 32-bit long). If an overflow occurs, you want to show an error message in *Mathematica* and then return the symbol **$Failed** instead of the sum.

You can use MLEvaluate to trigger the message, but how do you teach *Mathematica* about the message in the first place? You use an :Evaluate: line in your .tm file:

```
:Evaluate:  AddTwo::ovflw = "The sum cannot fit into a
C long type."
```

The addtwo function now looks like this:

```
void addtwo (void)
{
  long i, j, sum;
```

```
    MLGetLongInteger(stdlink, &i);
    MLGetLongInteger(stdlink, &j);
    sum = i + j;
    if(i>0 && j>0 && sum<0 || i<0 && j<0 && sum>0)
    {
        /* integer overflow usually shows up as
           "wrapping" to the other end of the range */
        MLEvaluate(stdlink, "Message[AddTwo::ovflw]");
        MLNextPacket(stdlink);
        MLNewPacket(stdlink);
        MLPutSymbol(stdlink, "$Failed");
    }
    else
    {
        MLPutLongInteger(stdlink, sum);
    }
}
```

Try compiling this version of addtwo, installing it, and evaluating **AddTwo[2^31-1, 2]**. You will get the overflow message.

After the call to MLEvaluate, *Mathematica* will send back a Return-Packet containing the return value of the Message function (which is simply the symbol Null). You need to clear this packet from the link, so you call MLNextPacket (which will return RETURNPKT) and then call MLNew-Packet to discard the contents. If you wanted to read the contents of the ReturnPacket, then you would replace MLNewPacket with an appropriate series of MLGet calls. As an example, suppose you wanted to have *Mathematica* report its version information to you. Here is how you would send the request and read the result:

```
    MLEvaluate(stdlink, "$Version");
    MLNextPacket(stdlink); /* a RETURNPKT will be waiting */
    MLGetString(stdlink, &my_string);  /* inside there will
                                          be a string */
```

MLEvaluate is a convenience function which creates the expression EvaluatePacket[ToExpression["the string"]] and sends it to *Mathematica*. Whenever *Mathematica* calls your external function, it then initiates a read from the link and waits until the function returns a final result. EvaluatePacket indicates to *Mathematica* that the packet to follow is not the final answer. *Mathematica* will evaluate the packet, return the result, and keep waiting for the external function to return its final answer. In this way, the external function can initiate dialogs of arbitrary length and complexity with the kernel before it returns.

Simple requests can be sent as strings using MLEvaluate (especially when the exact request is known at compile time). It is often necessary, however, to construct the request using MLPut functions and to wrap the whole request in EvaluatePacket. For example, if you needed to invert a matrix as part of your external function, you would write:

```
MLPutFunction(stdlink, "EvaluatePacket", 1);
MLPutFunction(stdlink, "Inverse", 1);
MLPutDoubleArray(stdlink, tdata, tdimensions,
                 heads, 2);
MLEndPacket(stdlink);
```

where tdata,tdimensions, and heads are defined as in an earlier example. You read the resulting ReturnPacket using MLGetDoubleArray in the same way as before.

7.7   Handling and recovery of *MathLink* errors

Robust external functions require error checking. If the external function terminates abnormally due to careless programming, *Mathematica* will notice that the link has died (at best) and respond with a cascade of one or more warning messages. It is your responsibility to check for *MathLink* errors in your external function. Most *MathLink* functions return 0 to indicate that an error has occurred; you should, therefore, check the return values, especially for the MLGet functions. You then need to recover from any error and inform *MathLink* that you have cleared the error condition. If you continue to issue *MathLink* calls after an error has occurred without clearing the error, things will no longer work as expected. In particular, *MathLink* calls effectively become no-ops (they smile, nod their head, and ignore you). mprep-generated code does an excellent job of checking for MLGet errors for argument and return types other than Manual. If you do not have any Manual arguments, then you will not have to worry about error checking. We suggest you take a look at the mprep-generated code to see how error checking should be handled.

The exact series of steps you take after an error has been detected depends on whether you want to try to recover or not. If you do not want to attempt error recovery after an MLGet call fails, the easiest thing to do is to abandon the external function call completely and return the symbol **$Failed**. This allows *Mathematica* to proceed, but the user is not likely to know what happened in your external function, nor will the user know what remedy to pursue. It is far better to trigger a diagnostic message. There is a

MathLink function `MLErrorMessage` which returns a string describing the current *MathLink* error. Use this string to inform the user what went wrong. The following code fragment demonstrates how to detect a *MathLink* error, construct a useful message, and exit the function call. For each `MLGet`-type call in your code, you will have a construct such as:

```
if(!MLGetLongInteger(stdlink, &i)) /* for example */
{
  char err_msg[255];
  sprintf(err_msg, "%s\"%.76s\"%s",
          "Message[AddTwo::mlink,",
            MLErrorMessage(stdlink),
              "]");
  MLClearError(stdlink); /* turns MathLink on after
                            error*/
  MLNewPacket(stdlink);  /* clear out the remainder of
                            the packet */
  MLEvaluate(stdlink, err_msg); /* send the error
                                    msg */
  MLNextPacket(stdlink); /* look for the result of the
                            error msg */
  MLNewPacket(stdlink); /* throw away the packet */
  MLPutSymbol(stdlink, "$Failed"); /* NOW tell
                                      Mathematica */
  return;
}
```

Any of these *MathLink* calls might fail also! For example, consider the case where *Mathematica* is running on one machine and the external function is running on another machine. The machine running *Mathematica* suffers a power failure. Obviously, when the `MLGetLongInteger` call fails, it is not going to be helpful to try to post messages to the defunct *Mathematica*! However, this is better than no error checking at all. If you have more than one or two `MLGet` calls in your code, you would want to implement this as a function or macro.

This example assumes that the message triggered here, `AddTwo::mlink`, is defined in the `addtwo.tm` file as follows:

```
:Evaluate:  AddTwo::mlink = "There has been a low-level
MathLink error. The message is: `1`."
```

What will this look like to a user? In an earlier example, the `AddTwo::ovflw` error message is defined to handle the case of two valid integers whose sum overflows:

In:

```
AddTwo[2000000000, 1000000000]
```

Out:

```
AddTwo::ovflw: The sum cannot fit into a C long type.
$Failed
```

The new `AddTwo::mlink` error handles the case where one or both of the arguments sent by *Mathematica* cannot be read properly by `MLGetLongInteger` (for example, if the argument in question is too large to fit into a C `long` type):

In:

```
AddTwo[5000000000, 1]
```

Out:

```
AddTwo::mlink: There has been a low-level MathLink
                  error.
The message is: machine integer overflow.
$Failed
```

Many errors such as these are much more easily detected by *Mathematica*. How would you modify the `AddTwo` template (as opposed to the `addtwo.c` function) to prevent integer overflow? Hint: consider the *Mathematica* functions **N** and **NumberQ**. Checking validity in *Mathematica* is much easier!

7.8   Handling and recovery of other errors

After your first few *MathLink* programs, you will rarely make *MathLink*-related mistakes. However, there are many other sources of program runtime error besides *MathLink*, which is quite free of defects and easy to use. In particular, if you are using legacy code or libraries (for example, canned algorithms out of a book or a commercial library), then you need to handle error messages and defects introduced by your external code in a way that is correct and elegant to a *Mathematica* programmer. The following extended example shows how to call the algorithm `plgndr` from the very popular text *Numerical Recipes in C* by Press et al. (1992). Note that no changes need to be introduced into the original algorithm. All that is needed is a *MathLink* template along with judicious rewriting of the `nrutil` files listed in the index of that book. Here, we show how to use so-called "exception processing semantics" to propagate errors out of the bowels of a library up to a level at which they can be handled properly. You can apply similar techniques on other legacy libraries.

The `plgndr` function computes the associated Legendre function $P_l^m(x)$:

```
float plgndr (int l, int m, float x)
/* Here m and l are integers satisfying 0 <= m <= l,
   while -1 <= x <= 1 */
{
  void nrerror (char error_text[]);
  float fact, pll, pmm, pmmp1, somx2;
  int i, ll;

  if (m < 0 || m > 1 || fabs(x) > 1.0)
      nrerror("Bad arguments in routine plgndr");

  /* BUY THE BOOK IF YOU WANT TO SEE THE NIFTY CODE! */

}
```

Note that the algorithm checks for valid arguments and, if the test fails, calls `nrerror` (which resides with other utility functions in the files `nrutil.h` and `nrutil.c`). There is a problem here because the default behavior of `nrerror` is to call the C library function `fprintf`, which prints a string to the system console; it then calls the C library function `exit`, which terminates (with extreme prejudice!) the program. This is unacceptable in a *MathLink* template-based program for several reasons. First, on many systems (MacOS and Windows, for example) there is no system console – and therefore there is no place to print the message! Second, killing the external *MathLink* program causes the link between it and *Mathematica* to die. This generates the unfriendly message

```
LinkObject::linkd: LinkObject[myProg, 1, 1] is closed;
the connection is dead.
```

within the *Mathematica* notebook, which tells you little or nothing about the source of the error. This message is a symptom of the error, not the cause. Third, consider the following:

```
Plot[NRplgndr[3,2,x], {x,-2,2}]
```

Plot will call `NRplgndr` (a wrapper around `plgndr`) many times. The most *Mathematica*-like behavior would be to generate some error messages in the notebook and soldier on, generating a plot for the range **{x, -1, 1}** where `plgndr` does work. Last of all, and more subtly, many of the algorithms in *Numerical Recipes* dynamically allocate memory (using utility functions such as `dmatrix` for allocating matrices of doubles); unfortunately, they do not attempt to free the memory if an error occurs and `nrerror` is called. This so-called "barf and bail" strategy of handling errors is acceptable for a program that will die upon the first error, but it will cause a memory leak if used within a program that ought to recover gracefully.

Here is the challenge. With no changes to the original `plgndr`, we must figure out a way to recover gracefully upon error without leaking memory, and we must find a solution that applies unchanged to many if not all of the algorithms in *Numerical Recipes*. You cannot simply filter out incorrect inputs in *Mathematica* before invoking `plgndr`. Even though such a filter would work for this function, many other functions call `nrerror` at less predictable times (for example, when an iterative algorithm fails to converge). Think about this before continuing. Draw pictures. Look at *Numerical Recipes*. Press et al. (see page 2) state, "In some applications, you will want to modify `nrerror()` to do more sophisticated error handling, for example to transfer control somewhere else, with an error flag or error code set." We assure you that you will be faced with similar design problems whenever you want to integrate someone else's code and lack the freedom to modify that code.

If you have wrestled with this and are ready to continue, let us explore one reasonable solution to the problem. First of all, here is a template file `NRplgndr.tm`:

```
:Begin:
:Function:    NRplgndr
:Pattern:    NRplgndr[l_Integer, m_Integer, x_]
:Arguments:   { l, m, N[x] }
:ArgumentTypes: { Integer, Integer, Real }
:ReturnType: Manual
:End:

void NRplgndr(int l, int m, double x)
{
 float res;

 NR_TRY
  res = plgndr(l, m, x);
  MLPutFloat(stdlink, res);
 NR_RECOVER
  MLPutSymbol(stdlink, "$Failed");
 NR_ENDTRY
 return;
}
```

(Note: Template files can contain C code, which `mprep` passes unchanged to its output `.tm.c` file. Often it is useful to keep a template and a C function that wraps your legacy code in one file; they have a common purpose and so belong in a common file.)

Let us examine the template and wrapper file and see what is happening. First, NRplgndr takes the same arguments in the same order as plgndr. The template passes {l, m, N[x]} because we want to make sure that the third argument is converted to a floating-point number; the user might have called NRplgndr with an integer for its third argument.

Why do we use Manual for the ReturnType? We use Manual because we might not want to return a number; we might want to signal failure, which is expressed as **$Failed**. NR_TRY, NR_RECOVER, and NR_ENDTRY are macros we have defined in modified nrutil.h and nrutil.c files. They implement something known as "exception handling semantics." The NR_TRY clause means exactly that – to try it. We try to compute the unchanged plgndr function, returning the answer to *Mathematica* with the *MathLink* function MLPutFloat. If anything goes wrong in plgndr (or in anything called by plgndr), an exception is raised, and processing jumps to the NR_RECOVER clause, where we put the appropriate result for *Mathematica* (in this case signalling failure).

Exception processing is a powerful concept that allows you to avoid cluttering your code with endless if statements. You try code assuming it will work flawlessly and recover if something goes wrong. Exception handling is part of *Mathematica*'s design, as well as part of the C++ and Java languages, in which exceptions are first-class objects in the language. The older languages such as C and FORTRAN do not have exceptions; in the case of C it can be mimicked rather well with the standard C library functions setjmp and longjmp. These two functions allow one to define a jump point and later return to it. For clarity's sake, we hide much of the messy details in the C macros NR_TRY, NR_RECOVER, and NR_ENDTRY, defined as follows:

```
#include <setjmp.h>
extern jmp_buf NRjmp_buf;
extern int setjmpCalled;

/* convenience macros for error recovery - use as
   follows */
/* NR_TRY            */
/*   NRsomerecipe();       */
/*   MLPutxxxxx();        */
/*   MLPutxxxxx();        */
/*   ...          */
/* NR_RECOVER         */
/*   MLPutSymbol(stdlink, "$Failed"); */
/* NR_ENDTRY         */
```

```
#define NR_TRY if (!setjmp(NRjmp_buf)) { \
    setjmpCalled = 1;{
#define NR_RECOVER }} else {{ \
    MLClearError(stdlink);
#define NR_ENDTRY } setjmpCalled = 0; }
```

What happens when plgndr calls nrerror? The new definition of nrerror is as follows:

```
void nrerror(char error_text[])
/* Numerical Recipes standard error handler */
/* This handler has been modified to use MathLink to */
/* return its error message. It also returns $Failed */
/* and closes the link. */
{
 char errBuf[ERRBUFSIZE];
 sprintf(errBuf, "%s",error_text);
 msgv[0] = errBuf;
 send_message_to_mathematica("NumericalRecipes",
                             "nrerr", 1, msgv);
 free_plist(&NR_memorylist);
 if (setjmpCalled)
 {
  setjmpCalled = 0;
  longjmp(NRjmp_buf, 1);
 }
 else
{
  MLClearError(stdlink);
  MLPutSymbol(stdlink, "$Failed");
  MLEndPacket(stdlink);
  MLNewPacket(stdlink);
  MLClose(stdlink);
  exit(1);
 }
}
```

nrerror sends an error message to *Mathematica* (the moral equivalent to printing to a system console), frees any memory that has been allocated but otherwise abandoned by the algorithms in the book, and uses the function longjmp to jump over all intervening functions to the setjmp function call that is part of the NR_RECOVER macro. We do not discuss further the function free_plist, but we note the strategy behind it: the memory allocation functions in nrutil.c have been modified so that all blocks of memory so allocated are placed on a linked list. If nrerror is called, one can clean up the otherwise abandoned memory by walking through the list and properly deallocating it. This is not a *MathLink*-specific technique, but rather we are being explicit about the problems of using low-level languages such as C. *Mathematica* code does not need to allocate or deallocate memory; the *Mathematica* programmer rarely has to think about such issues. The C programmer is not so lucky. Our example is trying to express the care you need to take to integrate third-party libraries into *Mathematica* in a robust manner. Error recovery is a worthy goal to strive towards; the care you take will be rewarded by the smile on your user's face.

To continue our example, we look at the function that sends error messages to *Mathematica* from within C, a variant of the modified addtwo example given earlier in this chapter.

```
/************** error handling **************/
static int send_message_to_mathematica(char *symb,
char *tag, int msgc, char *msgv[])
{
  int pkt,i;

  MLPutFunction( stdlink, "EvaluatePacket", 1);
  MLPutFunction( stdlink, "Message", msgc+1);
  MLPutFunction( stdlink, "MessageName", 2);
   MLPutSymbol( stdlink, symb);
   MLPutString( stdlink, tag);
  for (i = 0; i < msgc; i++)
      MLPutString( stdlink,msgv[i]);
  MLEndPacket( stdlink);

  while( (pkt = MLNextPacket( stdlink)) && pkt !=
                                          RETURNPKT)
      MLNewPacket( stdlink);
  MLNewPacket( stdlink);

  return MLError( stdlink) == MLEOK;
}


/* ....other stuff intervenes... */
send_message_to_mathematica("NumericalRecipes",
  "nrerr", 1, msgv);
```

What does this code do? It simply takes the string that was passed into nrerror and constructs a *Mathematica* MessagePacket[] of the form

```
NumericalRecipes::nrerr = "Numerical Recipes run-time
error; the string passed into nrerror."
```

As in the earlier example, the C code eats the reply *Mathematica* generates in response to this packet. The last piece of this puzzle is a template file called NRMain.tm which contains template code that is shared among all the *Numerical Recipes* wrappers one might want to create. Here is NRMain.tm:

```
:Evaluate: NumericalRecipes::nrerr = "Numerical Recipes
run-time error; `1`."
:Evaluate: numberQ[x_] := NumberQ[N[x]]
:Evaluate: numMatrixQ[x_] := MatrixQ[x, numberQ]
:Evaluate: numVectorQ[x_] := VectorQ[x, numberQ]
:Evaluate: numSquareMatrixQ[x_] := Apply[SameQ,
Dimensions[x]] && numMatrixQ[x]


#define NRANSI
#include "NR.H"
#include "NRUTIL.H"
#include "mathlink.h"

int main(int argc, char *argv[])
{
        return MLMain(argc, argv);
}
```

We see the customary and familiar C main function as well as an :Evaluate: statement that defines the common prefix to the class of *Numerical Recipes* errors.

Let us see what happens when we build this program. We input NRMain.tm and NRplgndr.tm into the mprep tool. This results in a single NRMain.tm.c file. We compile that and compile plgndr.c along with any other numerical recipes required and the utility file nrutil.c. We link the whole lot together with the *MathLink* library and get a small executable that we can **Install** and use from within *Mathematica*. A sample run from the actual tool follows: install the template program, try a legal input (**1** >= **m**), try an illegal input (**1** < **m**), and then see if NRplgndr is still running:

In:

    1=Install["NRplgndr"]

Out:

    LinkObject[NRplgndr, 2, 2]

In:

    NRplgndr[3,2,0.5]

Out:

    5.625

In:

```
NRplgndr[2,3,0.5]
```

Out:

```
NumericalRecipes::nrerr: Numerical Recipes run-time
error; Bad arguments in routine plgndr.
$Failed
```

In:

```
Plot[NRplgndr[3,2,x],{x,-1,1}]
```

Out:

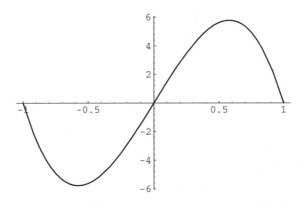

What happens if we plot over a range with legal and illegal values?

In:

```
Plot[NRplgndr[3,2,x],{x,-2,2}]
```

Out:

```
NumericalRecipes::nrerr:
    Numerical Recipes run-time error;
    Bad arguments in routine plgndr.
NumericalRecipes::nrerr:
    Numerical Recipes run-time error;
    Bad arguments in routine plgndr.
Plot::plnr: CompiledFunction[{x}, <<1>>, -CompiledCode-
][x] is not a machine-size real number at x = -2..
NumericalRecipes::nrerr:
    Numerical Recipes run-time error;
    Bad arguments in routine plgndr.
General::stop:
    Further output of NumericalRecipes::nrerr
        will be suppressed during this calculation.
Plot::plnr: CompiledFunction[{x}, <<1>>, -CompiledCode-
][x] is not a machine-size real number at x = -1.83333.
Plot::plnr: CompiledFunction[{x}, <<1>>, -CompiledCode-
```

```
][x] is not a machine-size real number at x = -1.66667.
General::stop:
    Further output of Plot::plnr will be suppressed
during this calculation.
```

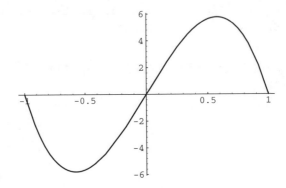

Note that the care we took ensures that NRplgndr, which is written in C, really behaves like a *Mathematica* function. We can use **Plot**, which repeatedly evaluates its argument with a very clever adaptive plotting algorithm. We survive the illegal *x*-values ($|x| > 1$) and plot the legal ones. This is about as complex as it gets. In a later chapter, we will wrap the functions of a data acquisition library to incorporate its functionality in *Mathematica*. If you consider intermittent errors in a data acquisition board to be analogous to errors in a numerical algorithm, you might see why the error handling techniques employed in this extended example apply to building a robust data logging system in *Mathematica*.

7.9   Troubleshooting and debugging

When you embark upon substantive *MathLink* development, you will be writing C code that is much less forgiving than *Mathematica*. Learn to use the symbolic (high-level) debugger that comes with your C compiler – it is your friend. In this section we will give tips to help you flush out errors.

If you get either one of these two typical errors when you try using **Install**

In:

```
    Install["myProg"]:
```

Out:

```
    LinkOpen::linkf: LinkOpen[myProg] failed.
    LinkConnect::linkc: LinkObject[myProg, 1, 1] is dead;
    attempt to connect failed.
```

then either the program is not being found, or else it is launching and then immediately crashing or quitting. If you **Install** a program that exists but is not properly *MathLink*-aware, then **Install** will hang until you abort it. **Install** does not interpret its argument, and in particular it does not search the directories on **$Path**. Its argument is simply passed to the operating system's or user's shell (such as the Macintosh Finder), which will search for myProg according to its own rules. On UNIX, for example, the path that is searched is the path inherited by shell processes launched by the kernel. You may need to give a complete pathname to the program. You can check your program by launching it from the command line (under UNIX) or by double-clicking it (on Macintosh or Windows machines). You should get a "Listen on:" prompt, which you can dismiss, followed by a "Connect to:" prompt, which you also dismiss, at which point the program will terminate normally.

If your program works correctly when you perform the above test but otherwise behaves strangely, you need to perform more extensive debugging. If the program crashes because of something in your non-*MathLink* code, or if it exits because you are using *MathLink* calls incorrectly, you will probably see the following message:

```
LinkObject::linkd: LinkObject[myProg, 1, 1] is closed;
the connection is dead.
```

If there is a simple error in your *MathLink* code, run the program using the symbolic debugger supplied with your compiler. Following are some guidelines on how to make your external program run under a debugger and still communicate with *Mathematica*.

To run your installable program with a symbolic debugger, you need to launch it from your debugger and then manually establish a connection with *Mathematica*. This issue is discussed in the *MathLink* documentation, along with an example using the UNIX gdb debugger. The details differ among debuggers, but the steps you need to follow are the same.

- Launch the program in your debugger.
- Set breakpoints wherever you suspect a problem.
- When you tell the debugger to run (or go) and your program tries to open a link, it will give a "Listen On:" prompt.
- Supply any arbitrary link name you prefer. Macintosh and Windows systems use arbitrary strings (for example, myBuggyProgram); on UNIX, you need to use an unused port number (for example, 3000).
- Now, switch to *Mathematica* and type:

```
link = Install["name", LinkMode->Connect]
```

where **name** is the linkname you specified in the "`Listen On:`" prompt. Use string quotes around this name, even if it is a number. Note that **Install** can take the same sort of arguments that LinkOpen takes. Here, we give a linkname as the first argument (when we want *Mathematica* to launch the program, this is just the filename), and we specify link options as well.

At this point, you should be able to step through your program in the debugger and see what is causing the problem.

7.10 Large projects

Most of our examples so far have all been single functions. The potential for installable functions is much greater, though; you can write complex installable programs that provide complete solutions for specialized problems. We need to discuss some issues that arise when constructing complex installable programs.

A complete solution to a problem usually requires the ability to handle many different types of input. Because it is tedious to write general interpreters in C or similar languages, and because *Mathematica* excels at pattern matching, you will want to write *Mathematica* code to go along with your C functions. Writing wrapper functions in *Mathematica* that process the options, arguments, and errors (thus "preflighting" the inputs to your compiled code) will make your C code much simpler. Make these wrapper functions visible to the user. The wrappers then call private functions (which are named in templates) that map directly to functions in the external program. This convention will allow you to evolve the *Mathematica*-visible names separately from the actual implementations in the template program.

Through the use of :Evaluate: lines in your .tm file, you can create an entire *Mathematica* package in your program source files; there is no separate .m file to load. The advantages of this are the convenience for users (they can just **Install** the program and be ready to go) and the additional concealment of any proprietary information in your program. The only disadvantage is that any modification of your package code requires you to reprocess the template with mprep and recompile the resulting C code.

You should decide whether you want *Mathematica* package code embedded in the external program (such that the user types **Install["prog-name"]** to use it) or want to have a package (.m) file that calls **Install** within it (such that the user types **<<Packagename`**). Because **Install** does not respect *Mathematica*'s **$Path**, the latter approach requires users either to give the program a predetermined name and place it where it will be

found automatically by **Install**, or to edit the .m file to reflect the user's preferred name and location of the program. This latter approach properly supports *Mathematica's* context-handling functions so that **Get** and **Needs** work with your package just as with any other package. Choose your preferred method.

If you plan to distribute source code for your program, you can consider embedding C code directly in your .tm file; it will be passed unchanged through mprep. This means that you do not need separate .c and .tm files. You then will have a single, self-contained file. Here is a sample of the suggested structure of such a .tm file:

```
:Evaluate:    BeginPackage["MyPets"]

    Put all of the package code here, in :Evaluate:
sections...

:Evaluate:    MyDog::usage = "MyDog has fleas."
:Evaluate:    MyCat::usage = "MyCat has hairballs."

        etc....

    The C code begins:

#include "mathlink.h"

void private_dog() { ...
void private_cat() { ...

        etc....

    Templates begin:

:Evaluate:    Begin["MyPets`Private`"]

:Begin:
:Function:    private_dog
:Function:    private_cat

        etc....

:Evaluate:    End[ ]
:Evaluate:    EndPackage[ ]
```

A commercial-quality installable program should behave as if it were a built-in package. Make sure that your functions are abortable and that they return

Mathematica-style messages for all errors or warning conditions. See the above example from *Numerical Recipes* for hints on how to do this.

7.11 Special topics

7.11.1 Unknown result length from your program

`MLPutFunction` requires you to state an exact number of arguments. Sometimes you do not know how many arguments you will send. For example, a particle physicist might want to acquire bubble chamber events for a fixed time period. She does not know how many events are going to be generated. There are a couple of approaches to avoid this difficulty.

The straightforward method does not require any special *MathLink* programming at all. Allocate enough local storage in your C program to hold the complete answer, counting answer components as you store them. When you finish, put all the data on the link at the same time. This is easy, except that you need to know how to perform dynamic memory management, which is a source of many a programmer's late-night debugging sessions. You also may need to do a lot of allocating and reallocating of memory to hold the result as it grows. You also need to free it all before finishing. You could also have your C program stream data to disk. This is a useful approach for data loggers that accumulate more data than will fit in main RAM.

How could you send the elements as they are generated by your external program? If you prefer to gather all the results and send them to *Mathematica* in one step, but you have no desire to brave the troubled waters of dynamic memory management, then there is another method you can use involving the use of a loopback link. This method is the most elegant and probably the most desirable, except in cases where the speed of the *MathLink* transfer is the most important consideration. Think of a loopback link as a way of getting the *MathLink* library to handle dynamic memory allocation for you.

7.11.1.1 Loopback links

Version 2.2 of *Mathematica* introduced a new link mode: the loopback mode. (Legal values of `LinkMode` are `Launch`, `Listen`, `Connect`, and `Loopback`.) This link type is quite useful, but it is unknown to most *MathLink* programmers since only brief documentation exists in the Major New Features of *Mathematica* Version 2.2 document.

A loopback link derives its name from the diagnostic mode of communication devices such as modems and terminals. In loopback mode, such devices tie their output to their input. Similarly, a loopback link is a link that loops back at you. You both write to it and read from it. A loopback link acts as a FIFO (first in, first out) buffer that gives the C programmer a *Mathematica* expression type.

Among the interesting uses of loopback links is a solution to the problem discussed in the previous section: how to send a compound expression to *Mathematica* if you can not call MLPutFunction because you do not know in advance the third argument. A loopback link provides a very simple solution: generate the elements of the result, put them on a loopback link, and count them as you go. *MathLink* will slay the dragon of dynamic memory allocation for you. When you are ready to send the result to *Mathematica*, you can use your stored count as the third argument in MLPutFunction.

Error handling, especially when the appropriate response to the error is to send no data to *Mathematica* (except for gracefully failing or aborting), is much easier with loopback links. Do not send anything to the kernel until the computation is finished. At that time you can decide to send whatever you want.

To open a loopback link, specify loopback as the link mode. For example, here is a function that returns a list of integers to *Mathematica* by first placing them on a loopback link.

```c
void BatchAcquire(int seconds)
{
  int i, sample_count;
  char loopback_argv[3] = {"-linkmode", "loopback",
                           NULL};
  MLINK samples;

  samples = MLOpen(2, loopback_argv);
  if(!samples)
  {
  /* might want to issue a message as well */
    MLPutSymbol(stdlink, "$Failed");
    return;
  }

  sample_count = 0;
  while(some_test_involving_clock_and_
          data_acq_hardware_working)
  {
    i = AcquireSampleFromBoard();
    MLPutInteger(loopback, i);
    sample_count++;
  }
```

```
        MLPutFunction(stdlink, "List", sample_count);
        for(i=1; i<=sample_count; i++)
            MLTransferExpression(stdlink, loopback);

        MLClose(loopback);
    }
```

`MLTransferExpression` is described in the Major New Features of Version 2.2 document. It provides a very convenient means for moving expressions from one link to another, since you do not need to be concerned with the exact structure of each expression. The destination link is given first, the source link second.

Note that *MathLink* handles the memory-management issues. There are no calls to `malloc`, `realloc`, or `free`, (or `NewHandle` or `NewPtr` for Macintosh folks who choose not to use the ANSI C library functions). You also do not have to worry about writing past the end of your storage and stomping over another data structure (or program code). On the other hand, you still need to check for errors in your `MLPut` calls if you are storing a lot of data since you could exhaust available memory, in which case *MathLink* will return an error when you attempt to `MLPut` another datum.

7.11.2 Making your function abortable

It is important for well-written software to give the user at least the illusion of control. One important aspect is allowing the user to abort or quit whenever wanted. A significant (and ongoing) effort was made in *Mathematica* to ensure that, when the user wants to stop, the program stops. If your function takes significant time to execute, you need to make it abortable. That is, when the user types the usual abort key sequence (for example, Control-C in UNIX or Command-period on Macintosh), the function should terminate as quickly as possible and return something appropriate. To do this you need to know two things: *i*) How does my program know when the user wants to abort the current operation?, and *ii*) What should it send back to *Mathematica* to acknowledge the user's request?

In an analogous fashion to RS-232 serial communication, a link contains two separate channels of communication (think of data and control information). One channel is for the *Mathematica* expressions; the other is for urgent messages that need to be sent in the midst of communication. The two salient examples are requests to interrupt or abort execution. This second channel is the one that is managed by the `Message` functions in *MathLink* (for example, `MLPutMessage`, `MLGetMessage`). Do not confuse these with the `MLErrorMessage` function or the familiar *Mathematica* error messages. The similarity in names is an unfortunate historical accident.

Normally, you do not need to worry about this at all. Handling messages from the kernel is performed in mprep-generated code. Suffice it to say that there is a global variable MLAbort in installable programs whose value will be set when the user wants to abort. This answers our first question (at least for computers with preemptive multitasking – see below for computers without preemption). When performing a lengthy calculation or other process, your code should periodically check the value of MLAbort. If it is nonzero, then you know it is time to stop the current operation.

What about the second question? What should your function return to *Mathematica* when the user aborts the evaluation? The *MathLink* Reference Guide suggests sending the symbol **$Aborted**, which is what your function will return when you do use a return type other than Manual in your template; in this case the mprep-generated code sends the final answer to *Mathematica*. That code checks the variable MLAbort, and if it is set, **$Aborted** is sent no matter what your function returns.

This is probably not the ideal behavior. If you are writing a *Mathematica* function that uses external functions, you want the whole expression to evaluate to **$Aborted**. You do not want to require all functions that use your installable function (and functions that use such functions *ad nauseam*) to handle intermediate values of **$Aborted**. You must propagate any abort request (as indicated by MLAbort) back to *Mathematica* in such a way that the abort continues to be processed. Thus, instead of sending the symbol **$Aborted**, send the function **Abort[]**. *Mathematica* will then halt the entire evaluation no matter how deeply nested your external installable functions are. This behavior mimics pure *Mathematica* programs.

As a last detail, programs running on systems without preemptive multitasking (for example, MacOS and Windows) need to yield the processor so that the *Mathematica* kernel has a chance to send the abort message. The easiest solution (that will also keep your code portable) is to call the yield function that mprep generates for template programs; this function will yield the processor temporarily to other programs.

Since many *MathLink* functions call this yield function internally (especially the Put and Get calls) you do not need to worry if you are making MLPut calls periodically during your computation. Call the yield function explicitly only if your function has long computation loops in which it does not make any MLPut calls before checking the value of MLAbort. You also do not need to do it if you are running under Unix (but you might want to, for portability reasons).

Here is a skeleton of a template program that performs a long calculation and periodically checks MLAbort. MLCallYieldFunction was introduced in Version 2.2.2; do not worry about the arguments, and use it exactly as it appears below.

```
void AcquireAtASlowRate(void)
{
  int result;
  int statusbit = 1;

  while(statusbit && !MLAbort)
  {
    statusbit = acquire_sample(slow_rate, &result);

    MLCallYieldFunction(MLYieldFunction(stdlink),
      stdlink, (MLYieldParameters)0 );
  }

  if(MLAbort)
  {
    MLPutFunction(stdlink, "Abort", 0);
  }
  else
  {
    MLPutInteger(stdlink, result);
  }
}
```

To summarize, you should periodically check the value of MLAbort during time-consuming processing. Remember to call the yield function periodically so that the kernel process can send any abort message to you. If you are sending results to *Mathematica* manually, you should send the function **Abort[]** and return when you realize that it is time to abort. If you are not using Manual in the :ReturnType: line of your template file, you should abort by immediately returning any value from your function (since it will never be sent to *Mathematica*). The mprep-generated code will send **$Aborted** in its place. Finally, if you are in the midst of sending part of the result when you detect the abort, call MLEndPacket prematurely to cause *Mathematica* to get the symbol **$Aborted** by default.

7.12   References

Janhunen, P., Stein, D., "MathHDF: A *MathLink*-based interface between *Mathematica* and HDF files for Distributed Visualization," Computers in Physics, May-June 1993.

Press, W. H., Flannery, B. P., Teukolsky, S. A., Vetterling, W. T., "Numerical Recipes in C (second edition)," Cambridge University Press, Cambridge, United Kingdom, 1992.

CHAPTER 8

Interfacing I: a simple serial link

Computers invariably work with numbers in a parallel format. For example, usually a byte is thought of as a group of 8 bits that are all present at some instant at some location, that is, not presented one at a time to a processor. Parallelism is the most time-efficient way to process data, and laying out parallel functionality on a silicon chip is straightforward. However, a serial format is convenient whenever some time efficiency can be sacrificed to gain simplicity and/or efficiency in some other aspect of the system.

The RS-232 standard defines a simple hardware protocol for serial data exchange between two pieces of equipment. In its simplest form, the link need consist only of three wires: a common or ground line, a transmit line, and a receive line. The three-wire cable is an advantage where space for cabling, connectors, or components within equipment is limited. We discuss aspects of the conversion from parallel data to serial data later because it is germane to parameters that have to be set by the software you will use.

For laboratory use, the RS-232 standard provides a cheap and simple solution to many interconnection problems because most computers are provided with a serial port that supports RS-232, to enable your computer to control both your experiment and the data acquisition process. Some laboratory instruments are controllable directly with RS-232; where data rates lower than ~20 kbit/s are used, RS-232–to–IEEE-488 converters can be used to provide a bridge between serial and parallel systems. For a homemade instrument system with an embedded microprocessor, the RS-232 line provides a natural way to communicate with the processor during system programming. In addition, once the instrument system becomes operational, the RS-232 line can be used both to control it and to upload or download data.

The basic RS-232 standard applies for distances up to approximately 15 m and at speeds up to approximately 20 kbit/s. The exact distance–speed combination will depend on cable type and equipment characteristics. For long distances (perhaps some kilometers) at higher bit rates, the link can be implemented using off-the-shelf twin fiberoptic cable, with light emitting diodes (LEDs) and photodiodes as the electro-optic transducers. The newer serial standards RS-422 and RS-485 employ differential circuitry that is more sensitive and has better noise immunity; cable lengths up to 1200 m and data rates up to 10 Mbit/s can be used.

In this chapter, we show you how you can control the serial port on both an Apple Macintosh and an IBM-campatible PC running Linux, a free UNIX-based operating system. For the Macintosh, our project has four distinct tasks: *i*) to make up and/or check the cabling required, *ii*) to write some software in C to drive the serial port, *iii*) to make the serial port C software work within *MathLink* by defining a template file, and *iv*) to check out our newly created functions from within *Mathematica*. For the Linux PC, tasks *ii*) and *iii*) are replaced by an edit of the system files that control how the serial ports are allocated and controlled.

8.1 The serial port hardware

To use the serial port, there are two aspects of the port's hardware about which we need to know: how to connect (electrically) the port to the outside world, and how to specify the necessary parameters to the hardware that converts the parallel data from the computer's internal bus into serial data.

8.1.1 Electrical connections

For this example, we use the minimum RS-232 configuration, which uses only the signal ground, transmit, and receive wires. The other connections specified in the RS-232 standard are used for controlling modems and are not necessary for simple instrument-control applications. Although we describe the computer-end connections so that you can create your own cabling, you will probably find that computer-to-modem cables, sold at your local electronics store, will do just fine.

The Macintosh serial port uses an 8-pin mini-DIN connector, with connections as defined in Table 8:i and shown in Fig. 8a; the PC uses the more common 9-way D-type connector with pin numbering adjacent to the pins and with connections as defined in Table 8:ii.

Although we have shown the connections for a Macintosh and the now-standard 9-pin PC system, if you are using a different type of computer (perhaps even a Macintosh with a different type of connector, since some older Macintosh machines had a 9-pin D-type connector), then you should check the computer's manual to confirm how its port is wired.

You also need to know how to connect the cable that runs from the serial port to the equipment to which you intend to connect. Although connecting just three wires between two pieces of equipment should be simple, many

Table 8:i   Macintosh serial port connections

Pin	Function
1	Handshake out
2	Handshake in
3	Transmit data −ve
4	Signal ground
5	Receive data −ve
6	Transmit data +ve
7	General purpose input
8	Receive data +ve

Table 8:ii   PC serial port connections

Pin	Function
1	Data carrier detect
2	Receive data
3	Transmit data
4	Data terminal ready
5	Signal ground
6	Data set ready
7	Request to send
8	Clear to send
9	Ring indicator

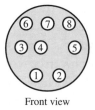

Front view

Figure 8a   Macintosh 8-pin DIN serial connector

equipment manufacturers' literature is opaque where it discusses which ends of the RS-232 link are treated as the communications end or the terminal end. Originally, the standard was intended for communications between modems and computers or between remotely sited terminals; if this connection model does not map onto your system then failing to work out which end is which can cause, say, both transmit lines to be connected, rather than the transmit to the receive. (It is rare for such a misconnection to cause a hardware failure, but the connection certainly will not work.) In addition to the polarization problem, some commercially produced RS-232 cables swap the transmit and receive lines and others do not! If you believe that you have a connection problem, we recommend *i*) that you check the continuity paths of all cables to establish that the transmit and receive wires are correctly paired and *ii*) that you use an existing package such as a terminal emulator to exercise the serial port.

The transmit pin on the computer port should be connected to the receive pin on the other equipment, and the computer's receive pin should connect to the other equipment's transmit pin. If you will work often with RS-232 connections, it is well worth buying line-checking equipment that has indicator lights to show you which line is active at any time. Such checking equipment is available at modest cost from most computer and electronic shops and mail-order catalogs.

For a quick look at what is happening, it can be helpful to use both some software that you know drives the serial port – such as a modem control package or a terminal emulator package – and an oscilloscope. Set the time axis to 100 μs per *x*-division, set the triggering to "automatic" with trigger selection to AC, and set the displayed channel to 5 Volts per *y*-division with the channel amplifier DC-connected. Adjust the intensity and the focus to get a sharp, easily visible trace. (The trace's *y* position can be changed by adjusting the position adjustment for the *y*-amplifier in use. With the triggering set to automatic, the trace should be continuous.) By connecting the oscilloscope probe's ground to the RS-232 signal ground and the probe tip to the transmit pin, you will be able to see the transmitted characters as a changing voltage level with one bit of data taking approximately one *x*-division square at 9600 baud (we discuss baud rate in the next section) and 100 μs per *x*-division. If you are using a higher baud rate, use a faster timebase setting or vice versa.

To gain familiarity, we suggest that you try looking at the RS-232 port of equipment that you know to be working. For further reading about the hardware aspects of RS-232, we have found the books by Frisch (1995), Horowitz & Hill (1980), and Thompson & Kuckes (1989) to be helpful.

8.1.2 Serial port parameters

When two pieces of equipment are connected together, each has to know how communication between them will be carried out. For serial transmission of bits down a cable, the simplest method is to represent the "1" and "0" bit values by different line voltages and, sequentially for all the bits in a word, to set the line voltage according to the bit value being transmitted. Without going into detail about the electronics required to execute the com-

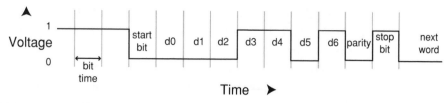

Figure 8b   Voltage vs. time for ASCII X

munication, we need to look at some of the aspects of serial transmission so that you have some understanding for the port parameters that you set in the software.

- The speed with which the bits are dispatched has to be known; otherwise, the receiving equipment cannot differentiate between a slowly transmitted 1 and two adjacent 1s transmitted quickly. The speed is set identically at both ends of the line in terms of the baud rate, a measure of the number of bits per second transmitted.
- The number of bits per word sent has to be specified. Typically 7 or 8 bits are sent per serial word.
- The end of a transmitted word is indicated by a 1 signal, called the *stop bit*. The duration of this stop bit has to be specified, because the RS-232 standard allows it to be 1, 1.5, or 2 times the length of a data bit. (The start of a word has a zero-valued stop bit.)
- A simple form of error-checking is often used: an extra bit (the *parity bit*) is optionally included at the end of a serial word and set so that the number of 1s in the word is even (for "even" parity) or odd (for "odd" parity).

The electrical levels and their timing are handled by special integrated circuits called either *universal asynchronous receiver-transmitters* (UARTs) or *asynchronous communications interface adapters* (ACIAs). It is the UART's job to convert the data, which is presented to it in parallel format on the computer's bus, into a serial format that can be transmitted down an electrical cable, one bit at a time. Figure 8b shows how the voltage varies as a function of time when an RS-232 line carries the ASCII character X. The bit time is the reciprocal of the baud rate; for example, a 9600-baud line will have a bit time of 104 μs, for example. (The voltage polarity may be reversed depending on where you examine the line, but this does not alter the fundamental characteristics of the transmission.)

To use a UART, the CPU has to load into it both the data to be sent and the control bits that specify to the UART's internal circuitry how the data should be transmitted. Similarly, at the other end of the cable, the receiving UART has to be controlled identically so that it can receive the serial data and convert them into a parallel format for the receiving processor.

Whichever computer system you are using, it is unlikely that you will have to look up the UART manufacturer's data sheet in order to see how to control the UART. Most operating systems and compilers come with driver software that will take care of the details of UART driving, leaving you free to keep your software at a higher, more application centered level. For example, the Symantec software that we shall use allows us to drive the Macintosh's serial port without even knowing what type of ACIA chip is used. If you want to know more about UARTs, we suggest you obtain a manufacturer's data sheet for a UART or ACIA chip; Garland (1979), Greenfield & Wray (1981), Thompson & Kuckes (1989) and also give useful information.

8.2   Software specification

Our task of software design can be started "top down" of "bottom up." With items of hardware that have limited functionality, it is necessary to design the lowest level of software "bottom up" because there is no point in reaching the end of a top-down design only to find that the base functionality that you have specified is unimplementable using the available hardware.

We need only four functions: one function each to open and to close the serial port, and one more function each to transmit to and to receive from the port one character or byte. These four functions can then be called from within *Mathematica* as often as we want so that arbitrary-length sequences of bytes can be transfered via the serial port. We could develop more complicated functions to handle more than one byte at a time, but many laboratory instruments require only a few bytes to be communicated and so the overhead of byte-at-a-time communication is minimal. If your application requires faster data handling, then you can develop more sophisticated functions by building on the functions described here.

The Macintosh-based example is explained in some depth; even if you are intending to implement the serial link on another machine, you will probably find the material it covers useful.

8.3   The Macintosh software

For this example, we use the Symantec C compiler, version 7. Although this compiler supports C++, we use only its C capabilities here. If you are not familiar with the compiler, we recommend that you take some time to read through its documentation and try out the programming examples described there so that you are familiar with creating a project, writing source code, running a program, debugging, and building an application.

8.3.1 Creating the rs232.π project

Before we can proceed to write any code for the serial link project, the Symantec compiler requires us to create a project document (with a name ending in .π) in which the compiler keeps track of which files are necessary for the project to function.

It is useful to keep the files required for one project in its own folder. We recommend that you begin by creating a folder called, say, `rs232 folder`.

Make a copy of the file `mprep.rsrc` (in the *MathLink* Resources folder in the *MathLink* development kit folder supplied with *Mathematica*). Rename your copy `rs232.π.rsrc`, and move it into `rs232 folder`.

Next, start up the `THINK C Project Manager` application. When the application starts, it will ask you to specify on which project you want to work. Create a new project, setting the type of project to be an `Empty project`. Name the project `rs232.π` and save it in the `rs232 folder` that you created earlier.

For our project to work, we need to add not only the application-specific source code, but also some system, C, and *Mathematica* libraries. Use the `Source/Add files...` dialog *i*) to add `AppleTalk`, `nAppleTalk`, `Mac-Traps`, and `MacTraps2`, all from the `Mac libraries` folder supplied with the compiler; *ii*) to add `ANSI` from the `Standard libraries` folder supplied with the compiler; and *iii*) to add `MathLink1.lib` and `MathLink2.lib` supplied in the `MathLink for THINK C` folder in the `MathLink Dev Kit` folder supplied with *Mathematica*. After you have added these files to the project, the next task is to write the serial port driver functions.

8.3.2 The application C code

In this section, we walk through the C code that we need to write for the four elementary functions necessary for serial port usage. To create a new source-code file, choose *New* from the compiler's *File* menu and add the code as described below.

C programmers will note that the code has no main section – that the code by itself, is not capable of being a stand-alone program. The code in this section will be used as functions by a program that will be created for us, as described in the next section.

The first part of the program contains instructions to the compiler to include a number of files that contain declarations for constants, variables, and functions. Filenames enclosed in angle brackets, < >, are system or compiler related files; the "mathlink.h" file is supplied with *Mathematica*. The constants OUTDRIVER and INDRIVER hold the system names for the modem port driver. The structure type SerStaRec is defined within the Macintosh toolbox and holds information about the status of the serial port.

```
#include <stdio.h>
#include <console.h>
#include <time.h>
#include <Serial.h>
#include <stddef.h>
#include <string.h>
#include <Types.h>

#include "mathlink.h"

#define XONCHAR 0x11
#define XOFFCHAR 0x13

#define OUTDRIVER "\p.AOut"
#define INDRIVER "\p.AIn"

#define SERBUFSIZ 1000

/* function prototypes */

int SerialInitialize();
int SerialWriteChar();
int SerialReadChar();
int SerialPortClose();

short inRefNum, outRefNum;
SerStaRec serialStatus;
```

The next part of code contains the source for the four functions that we shall use to control the serial port.

The first function is used to allocate the serial port to this application by opening the serial port device drivers, to set the transmission parameters for the input and output ports, to assign a buffer in which incoming data can be placed, and to define the handshake signals that will be recognized and used by the driver to control the bidirectional flow of data through the serial port. The functions OpenDriver, SerReset, SerSetBuf, NewPtr,

MemError, and SerHShake are all system functions from the Macintosh Toolbox. The type SetShk is a toolbox data structure that holds the definitions of the protocol used for serial port flow control; we have chosen to use the standard *xon/xoff* type of control whereby each end of the RS-232 link sends the characters control-S and control-Q to stop and start the flow of data on the link.

The call to SerReset defines how the transmission and reception of data will take place. SerReset sets the link speed in baud, the length of data words (either 7 or 8 bits long), the length of the stop bit at the end of each word (stop10 gives a one-unit-length stop bit, stop15 would specify a length of one and a half units), and the type of parity checking to be used (if required). The variables used to define the type of checking (for example, noParity) have values that are defined in the file Serial.h which is supplied with the Symantec compiler. If you want to change any of the port parameters, look in Serial.h to see what alternatives are available.

```
/******************************************///
SerialInitialize
int SerialInitialize()

/* This routine opens the modem-port driver and sets it
up for the baud rates, parity, stopbits, and word
length as required. */

{
  SerShk flags;
  Ptr buf;
  long count;
  short int err;
  inRefNum=0;
  outRefNum=0;

/* Open Serial Drivers */

  if ((err = OpenDriver(OUTDRIVER,&outRefNum)))
    return err;
  if ((err = OpenDriver(INDRIVER, &inRefNum)))
    return err;

/* Reset input and output drivers, and tx/rx
parameters. If any of these parameters need to be
changed, look in  Serial.h (it's part of the Symantec
compiler's files  (Apple #includes folder) for suitable
alternative names (which are all  much  what you'd expect
```

```
(e.g. baud9600)). */

    if ((err = SerReset(outRefNum, baud2400 + data7 +
                        stop10 + noParity)))
        return err;
    if ((err = SerReset(inRefNum, baud2400 + data7 +
                        stop10 + noParity)))
        return err;

/* Assign a SERBUFSIZ buffer for the input */

    if(!(buf = NewPtr(SERBUFSIZ)))
        return MemError();
    if (err = SerSetBuf(inRefNum, buf, SERBUFSIZ))
        return err;

/* Set handshaking for the input and output drivers */

    flags.fXOn = TRUE;
    flags.fInX = TRUE;
    flags.xOn = XONCHAR;
    flags.xOff = XOFFCHAR;
    if (( err = SerHShake(inRefNum,&flags)))
        return err;
    if (( err = SerHShake(outRefNum,&flags)))
        return err;

    return 0;
}
```

After the serial port has been initialized, we need a minimum of three more functions: one to write an 8-bit number to the port, another to read an 8-bit number from the port, and one to close the port so that it may be used by other applications.

The write function takes as its argument the number to be written. The C language does not have an 8-bit number type, so we declare the number as a short (16-bit) integer inVal, although we use only that number's least-significant byte. The Macintosh system toolbox contains the function FSWrite which can be used to transmit data to a file or device driver connected to a port. FSWrite takes three arguments: the driver reference number outRefNum (obtained during serial-port initialization), the address of a long integer num containing the number of bytes to transmit, and the address of the start of the data to be transmitted outbuf. Before inVal is passed to the system function FSWrite which will take care of handling the transmission, we need to shift the contents of the least-significant byte into the most-significant byte. (On Macintosh machines, addresses point to the most-significant byte of an integer, therefore, if we want to send one byte, that byte must be the one pointed to by the address passed to FSWrite. By moving the least significant 8-bits of inVal into the most-significant byte, the data occupy the correct address.)

After FSWrite is called, we call the Macintosh toolbox function Ser-Status to check for any error messages resulting from our call of FSWrite. If either FSWrite or SerStatus return an error message (indicated by a nonzero value), then our function returns with that error code. If both FSWrite and SerStatus have worked correctly, then the value returned is the number of bytes written by FSWrite. When FSWrite is called, num contains the number of bytes to write, but upon return from FSWrite the variable contains the number of bytes written. (Note that it is num's address that is passed to FSWrite so that num's value may be altered by the function.)

```
/****************************************///
SerialWriteChar
int SerialWriteChar(short int inVal)
{
   long num;
   OSErr err;
   short int outbuf;

   num=1L;
   outbuf=(inVal<<8);

   err=FSWrite(outRefNum,   &num, &outbuf);
   if(err)return(err);

   err=SerStatus(outRefNum, &serialStatus);
   if(err)return(err);

   return((short)num);
}
```

The read function takes no arguments. It first calls the Macintosh Toolbox function SerGetBuf to establish the number of bytes (returned in count) that are waiting in the serial port's buffer to be collected. If no bytes are waiting, the function returns zero without calling FSRead. If some bytes are waiting to be read, FSRead is called with count set to 1 so that only one byte is read. The value of that byte is returned from FSRead in the 8-bit character variable ch and subsequently is converted to a short integer (ich) before being used as the return value from the function.

```
/**********************************************///
SerialReadChar
int SerialReadChar()

/* Read one character from the serial port input. If no
character is available then return a zero, else return
the ASCII for that character.

NB If FSRead is called when no data are available,
it'll  wait infinitely until data appear! */
{
   long count;
   char ch;
   short int ich;
   ch=0;

   SerGetBuf(inRefNum, &count);
   if(count>0)
   {
      count=1;
      (void) FSRead(inRefNum,&count,&ch);
      ich=(short int) ch;
      return(ich);
   }
   else
      return(0);
}
```

The last serial-port function is used to close the port drivers. Before it closes
each driver, it verifies that a reference number for that driver exists; a zero
reference implies that the driver is not loaded and so does not need to be
closed. If either call to CloseDriver (the system function) fails, Seri-
alPortClose returns a nonzero error code; if both calls to CloseDriver
succeed, then a zero code is returned.

```
/**********************************************///
SerialPortClose
int SerialPortClose()
{
   OSErr err;

if(inRefNum)
   err=CloseDriver(inRefNum);
if(err)return(err);
if(outRefNum)
```

```
   err=CloseDriver(outRefNum);
return(err);
}
```

Finally, we need to write a main function that will be used by *MathLink* to communicate with the rest of our code.

```
/***************************************************///
main

int main(argc, argv)
int argc;
char* argv[];
{
   return MLMain(argc, argv);
}
```

Once you have entered all the source code, save it as `rs232.c` in the `rs232 folder`. After you have saved the file, add it to the project by using the `Source/Add Files...` dialog.

Now that we have written the port-driving functions, we need to create a program that links these functions with a running session of *Mathematica*, using the Macintosh's built-in interprogram communication facilities.

8.3.3 The *MathLink* template file

Creating the program that will link the functions we have written with a *Mathematica* session is simple. All we need to write is a template file in which we declare what functions are contained within the application code, what arguments they take, and what objects they return. Both the arguments taken and the objects returned are defined in terms of *Mathematica* types (for example, `Integer` and `Real`). The template file is then used as input data by a codegenerator (supplied with *Mathematica*) that writes the code that will be the application program (containing our port-driving functions) that *Mathematica* will execute in order to drive the serial port.

The template file contains elementary information on each function, delimited by `:Begin:` and `:End:` markers. You can use the compiler's normal source code–creation facilities to generate the template file. When you have finished writing the template file, remember to save it as `rs232.tm`.

```
:Begin:
:Function:      SerialInitialize
:Pattern:       SerialInitialize[]
:Arguments:     { }
```

```
:ArgumentTypes:    { }
:ReturnType:       Integer
:End:


:Begin:
:Function:         SerialWriteChar
:Pattern:          SerialWriteChar[i_Integer]
:Arguments:        {  i }
:ArgumentTypes:    { Integer }
:ReturnType:       Integer
:End:


:Begin:
:Function:         SerialReadChar
:Pattern:          SerialReadChar[]
:Arguments:        {   }
:ArgumentTypes:    {   }
:ReturnType:       Integer
:End:


:Begin:
:Function:         SerialPortClose
:Pattern:          SerialPortClose[]
:Arguments:        { }
:ArgumentTypes:    { }
:ReturnType:       Integer
:End:
```

After you have created the template file and saved it as rs232.tm, you need to use the *MathLink* preprocessor SAmprep to generate the C code that will actually run and manage the communication between *Mathematica* and your *MathLink* application code. (At this point, you can either quit THINK C or, using MultiFinder, switch to the operating system.)

The preprocessor SAmprep is part of the software that is supplied with *Mathematica*, in the *MathLink* development kit folder. When you execute SAmprep (by double-clicking on its icon), it will present a file-open dialog. Using the dialog, locate and select the rs232.tm file that you have written, and click on the *mprep* button. Another file-open dialog will appear, allowing you to specify where the file created by SAmprep will be saved. Using this dialog, save the rs232.tm.c file in the project folder; SAmprep will automatically terminate after you have saved the file.

From within THINK C, add the file rs232.tm.c to the project rs232. π, completing the preparation phase.

8.3.4 Building the executable program

We now use THINK C to build the executable program. The project manager window, titled `rs232.π`, should contain the files `AppleTalk`, `nAppleTalk`, `MacTraps`, `MacTraps2`, `ANSI`, `rs232.c`, `rs232.tm.c`, `MathLink1.lib`, and `MathLink2.lib`, each in a separate segment. The order of the files within the project manager window is not important, nor is the segment number assigned to any particular file.

Set the project type to `Application` and the file type to `APPL`. The `Partition size` required is `400` (K), and the `Far code` and `Far data` options are required. The `SIZE` flags box must be `58E0`. You can leave the `Creator` box as `????`. You should set the *THINK Project Manager* and *THINK C* options to `Factory Settings`.

Now that the project parameters are all set, we can build the application program (`Project` menu `Build Application...`). Name the application as `rs232` and save it in the `rs232 folder`.

8.3.5 The *Mathematica* code

Once you have written the C code and built the `rs232` application program, using it from within *Mathematica* is straightforward. After starting your *Mathematica* session, you must first install the application. Installation may require two actions. First, you invoke the **Install** function in *Mathematica*. Second (and only if you have whitespace in the file path), prompted by *Mathematica*, you will need to use a file-opening dialog box to launch the application. The dialog box has all the normal functionality of a typical Macintosh file-open dialog box except that there is a *Launch* button instead of an *Open* button. Click on *Launch* to make your C-code application available to your current *Mathematica* session.

Here is a *Mathematica* command sequence that you might use to communicate with an RS-232 serial device. The **Install** function takes the file name of the application as its argument; here, the application executable image is called `rs232` and is in a folder called `rs232` folder on the disk `Q127 Alpha`. **Install** returns the identifier used by *Mathematica* for the *MathLink*.

In:
```
rsLink=Install["Q127 Alpha:rs232 folder:rsTest"]
```
Out:
```
LinkObject[rsTest, 2, 2]
```

Before any transactions can take place on the serial link, you must initialize the computer's serial port; after initialization, you can test what functions are available for this *MathLink* by using the `LinkPatterns` function.

In:

```
siR=SerialInitialize[]
```

Out:

    0

In:

```
LinkPatterns[rsLink]
```

Out:

    {SerialInitialize[], SerialWriteChar[i_Integer],
    SerialReadChar[], SerialPortClose[]}

To write a byte-sized number to the serial port, you use the `SerialWrite-Char` function with the integer for transmission as the function's argument. This function returns either the number of bytes that were transmitted (which should be one) or an error code (if transmission failed). Information about the meaning of error codes generated by the Macintosh operating system is provided within the volumes of *Inside Macintosh* from Apple (1995). Symantec's "THINK Reference" product provides an alternative, online reference to both error codes and the multitude of functions within the Macintosh programming toolbox.

In:

```
i=77;
swR=SerialWriteChar[i]
```

Out:

    1

Many RS-232 devices echo back to the transmitting system each character they receive. If you are using such a system, the first character that you read back using `SerialReadChar` will be a copy of what you sent.

In:

```
srR=SerialReadChar[]
```

Out:

    77

You should continue to read characters until `SerialReadChar` returns a zero, thus indicating that no more characters are in the transmitting equipment's output buffer. You might perform a complete read using a `While` clause that is executed after each transmitted character, to catch not only the echoed character but also any further response from the other equipment.

In the following example, we use our *MathLink* RS-232 functions to communicate with a small computer system that is running an interpreted

BASIC. With BASIC, we can ask for the answer to a simple arithmetic problem by preceding the problem's content by a **?**, BASIC's shorthand for its **PRINT** function. We place the whole of the text to be transmitted in one *Mathematica* string, which we process with the **ToCharacterCode** function to convert the ASCII characters into their integer codes, which we then store in **txList**. We also append a carriage-return character (decimal 13) to **txList** to indicate to the receiving computer that our input has come to an end and should be processed.

In:

```
txList=ToCharacterCode["? 3+4"];
AppendTo[txList,13]
```

Out:

```
{63, 32, 51, 43, 52, 13}
```

Once we have a list of ASCII numeric codes, we use **SerialWriteChar** to transmit them one at a time. After each character transmitted, we invoke **SerialReadChar** within a **While** loop to receive any incoming characters until **SerialReadChar** returns zero, indicating that no more characters are available from the input buffer. Once reception is complete, we apply the function **FromCharacterCode** to the list of incoming character codes, **rxList**, to convert them into printable ASCII characters. (Note that the responding computer emits quite a few space (˘) and linefeed characters to format its output.)

In:

```
rxList={};
For[i=1,i<=Length[txList],i++,
    swR=SerialWriteChar[txList[[i]]];
    srR=-1;
    While[srR!=0,
          srR=SerialReadChar[];
          AppendTo[rxList,srR];
          ]
    ];
FromCharacterCode[rxList]
```

Out:

```
>?˘ ˘3˘+˘4˘

7


>˘
```

When you have finished with the serial link, you should close it, freeing the port for use by another application. `SerialPortClose` requires no arguments and returns an error code equal to zero for success:

In:
```
scR=SerialPortClose[]
```
Out:
```
0
```

8.4   The PC/Linux software

The task of driving the serial port on a PC running the Linux operating system is somewhat different to the Macintosh case. On UNIX systems – and Linux is highly UNIX-compliant – transactions with devices (including ports) are modeled as normal file transactions. Indeed, the system directory `/dev` holds entries that look just like files if you list its contents with the `ls` command. Each entry is the logical name of some device that may be on the system. For example, `/dev/hda2` will be the second partition on the IDE (internal) hard disk, `/dev/fd0` will be the first floppy disk drive, and `/dev/ttyS0` will be the first serial port. Of course, these files are rather special in that they are actually links to physical devices and so are eligible to have properties attributed to them, just as the serial port in the Macintosh example had, say, its set baud rate and parity. These device files also may be links; on some systems, the device name `/dev/cua0` might be synonymous with the first serial terminal line, `/dev/ttyS0`.

The task of driving the serial port under Linux is split between the applications program, which sees the port as a file, and the operating system, which must configure that port file appropriately. The applications program can also alter the port's attributes, but only if the operating system has enabled use of the port.

8.4.1   Port configuration

Port enabling is controlled by entries in two system files: `/etc/inittab` and `/etc/gettydefs`. These files hold, respectively, information about how the system is to initialize ports (as well as other system-wide attributes) and some identifiers that are used to describe terminal lines and set their attributes.

The procedures for port attribute setting are not difficult. Indeed, you may find that you do not need to set up the port differently from its default condition. We found that one version of Linux already configured the serial port as a 9600-baud connection; another version required the addition of the line

```
S0:234:  :  ttyS0
```

to the file `/etc/inittab` before that port would work. On both versions of Linux that we tried, running the C program that changes the port baud rate left the port configured at that new baud rate after we terminated that program and used *Mathematica* alone.

Basically, you need to make sure that the serial port you want to use is configured as a dumb terminal without login prompts or other unwanted activity. Because of the various flavors of Linux, you will need to examine the `inittab` and `gettydefs` files in your system to see what you need to add. Your local Linux expert or system administrator may be able to help, and the books by Frish (1995) and Welsh & Kaufman (1995) are worth reading for many other system issues, as well as port configuring problems.

8.4.2 Port use

It is possible to use *Mathematica* itself to communicate with the serial port. After all, Linux treats the serial port as a file, and *Mathematica* can write to and read from files! To communicate with the port, we have to open the port's file for writing and reading and, of course, to close it after use. We can use the function **WriteString** to send one or more characters to the port.

In:

```
of=OpenWrite["/dev/ttyS0"]
```

Out:

```
OutputStream[/dev/ttyS0, 3]
```

In:

```
WriteString[of,"abcdef"]
```

In:

```
or=OpenRead["/dev/ttyS0"]
```

Out:

```
InputStream[/dev/ttyS0, 4]
```

To read characters received, we can use **Read**. Processing the number read by **FromCharacterCode** produces a string (whose quotes are not shown unless you specify **FullForm**).

In:

```
input=FromCharacterCode[Read[or, Byte]]
```

Out:

6

You can write your own C code for reading and writing to the port, and manipulation of the `ioctl` structure can change port attributes such as speed, parity, and word length.

```c
#include <stdio.h>
#include <fcntl.h>
#include <signal.h>
#include <ctype.h>
#include <sys/ioctl.h>
#include <termios.h>

/* declare functions */

void openFile();
void closeFile();
void writeOneChar(char c);
char readOneChar();

/* declare global variables */

short int ttyRefNum;
char myInBuf[1000];
struct termio myTIO;

/*MAIN */

main(argc,argv)
int argc;
char **argv;
{
  char myBuf;
  long int loop;

  openFile();

  writeOneChar('>');

  while(myBuf!='z')
    {
      myBuf=readOneChar();
      if(isprint((int)myBuf))
```

```
            printf("\nread %c\n",myBuf);
      }

   closeFile();

   printf("\ndone!\n\n");
}
```

The four functions are quite straightforward. We use the UNIX C functions open and close to connect to and disconnect from the port. When we open the port, we set its mode to "read-and-write" with the mode-value 2. To change the port's operating parameters, the terminal control parameters have to be read into the structure myTIO using an ioctl function and the TCGETA flag; after the parameters are altered, ioctl with TCSETA writes the parameters to the system. To see which parameters can be changed and which constants that you can use, look in termios.h. For example, to change the baud rate to 2400, we must first clear those bits (the least significant four bits) of myTIO.c_cflag that hold the port speed by bitwise ANDing them with hexadecimal FFF0 and then ORing in the constant for 2400 baud, B2400, as defined in termios.h.

```
void openFile()
{
   ttyRefNum=open("/dev/ttyS0",2);
   if(ttyRefNum<0)
      {
         printf("\nfailed to open port\n");
         exit(1);
      }
   ioctl(ttyRefNum,TCGETA,&myTIO);
   myTIO.c_cflag&=0xFFF0;
   myTIO.c_cflag|=B2400;
   ioctl(ttyRefNum,TCSETA,&myTIO);
}

void closeFile()
{
   close(ttyRefNum);
}

char readOneChar()
{
   char answer[1000];
   short int loop;
```

```
unsigned long int bufLength;
bufLength=20L;

read(ttyRefNum,&answer,bufLength);
printf("\n");
for(loop=0; loop<10; loop++)
{
  if(isprint(answer[loop]))
  {
    printf("%c",answer[loop]);
  }
  else
  {
    printf("_");
  }
  printf("\n");
}
return answer[0];
}

void writeOneChar(char myChar)
{
  unsigned long int bufLength;
  bufLength=1L;
  write(ttyRefNum,&myChar,bufLength);
}
```

8.5   Problems?

If you compiled your C software without any errors, well done! If not, check over the code carefully to sort out any syntax errors; the compiler is normally excellent at finding these for you. For less tractable errors – for example, when the compiler shows no errors but the code does not work – here are some pointers.

8.5.1   Checking the hardware

If the C software compiles and links without problems and/or the *Mathematica* software runs without reporting any errors, but if you cannot see any sign of communication between the two ends of the RS-232 line, then you should first suspect the hardware connections.

Check that the RS-232 cable is firmly connected. Check that the computer's transmit pin is indeed connected to the receive pin of the other equipment. You can normally tell which pin is transmit on the other equipment by making it send some data and watching (say, using an oscilloscope) for activity on one of the two connected pins. The active pin will be the transmit pin, which should be connected to the Macintosh receive pin.

Check that both ends of the RS-232 line are expecting data to be sent at the same baud rate, with the same word length, with the same length of stop bit, and with the same type of parity. (Regardless of the type of parity transmitted, the receiving equipment can also choose to ignore the parity bit.)

8.5.2 Debugging the C code

Although it is possible for problems to occur in the template file, that file's simplicity makes debugging a simple task. We find that a rigorous check of spelling and the types of both transmitted and received parameters will catch any problems.

The task of debugging the C functions is potentially less easy because of the greater complexity of the code. However, there are two aspects of debugging that lessen the task.

First, THINK C is supplied with a very good online debugger with which you can step through your program, line at a time, watching program flow patterns and how variables' values change. There is an extensive tutorial on debugger usage in the THINK C documentation; we recommend that you familiarize yourself with at least the elementary functionality of the debugger.

Second, the way in which we have designed rs232.c makes it possible to run the program as a stand-alone C program, without any involvement from *Mathematica*. To test rs232.c, remove rs232.tm.c from the project by selecting it in the project management window and then choosing *Remove* from the *Source* menu. Create a new main function for rs232.c with appropriate calls to each of the serial-port functions. For example, your revised main might look like:

```
main()
{
  short int testStatus, testValue;
    testStatus=SerialInitialize();
    testValue=77;
    testStatus=SerialWriteChar(testValue);
    testValue=SerialReadChar();
    testStatus=SerialPortClose();
}
```

To test each function, run `rs232.c` with the *Use Debugger* option in the *Project* menu checked on. You can then use the *Step* command in the debugger to step through the program one line at a time, using the *In* command to go into functions called, as appropriate. Use the debugger's data window to display variable values and to show how they change as your program executes.

Note that the Macintosh's receive buffer (here set to 1000 bytes) must be long enough to contain all the output from the transmitting device. Too short a buffer will cause parts of the received message to be lost.

8.5.3 Response times

Note that many computers and instruments may require some time to process commands sent to them and also to generate any response. If your system appears to miss data transmitted to it, or if it responds to a new command with the answer to a command issued previously, then you may need to insert a delay between command transmission and response reception.

To implement such a delay, you can use *Mathematica*'s **Pause** function. You can also implement your own delay routine as part of the C code. The Macintosh toolbox contains the function `Delay(t, &tEnd)` which is declared in `OSUtils.h` and takes two unsigned long integer arguments: `t`, the number of machine ticks (1/60 s) by which to delay, and the address of a variable, `tEnd`, into which the system time at the end of the delay will be placed.

8.6 References

Apple Computer Inc., "Inside Macintosh CD-ROM," Addison-Wesley, Reading, Massachusetts, USA, 1995.

Frisch, A., "Essential System Administration," O'Reilly & Associates, Sebastopol, California, USA, 1995.

Garland, H., "Introduction to Microprocessor System Design," McGraw-Hill Kogakusha, Tokyo, Japan, 1979.

Greenfield, J. D., Wray, W. C., "Using Microprocessors and Microcomputers: The 6800 Family," John Wiley & Sons, New York, USA, 1981.

Horowitz, P., Hill, W., "The Art of Electronics," Cambridge University Press, Cambridge, United Kingdom, 1980.

Thompson, B. G., Kuckes, A. F., "IBM-PC in the Laboratory," Cambridge University Press, Cambridge, United Kingdom, 1989.

Welsh, M., Kaufman, L., "Running Linux," O'Reilly & Associates, Sebastopol, California, USA, 1995.

CHAPTER 9

Interfacing II: more advanced links

In this chapter we focus on connecting *Mathematica* to more complicated laboratory hardware to make a *"Mathematica* workbench." We focus on two major examples. In the first, we build upon the material of the previous chapter and show you how to link *Mathematica* to a simple device that is "naked" – that is, lacking any special supporting software. In the second case, we show you how to incorporate a richly featured library that supports a hardware product family. You are likely to encounter either situation depending upon your laboratory budget and the scope of your task.

Before proceeding to these two case studies, we need to mention a few important issues that affect how you choose to connect *Mathematica* to a laboratory system. These issues are *layers, loops,* and *latency*.

9.1   Design issues

9.1.1   Layers

In Chapter 6, we described a laboratory data acquisition system in many different layers. As we move between these layers, the nature of the detail changes. Consider an electrocardiogram and a defibrillator. A doctor might describe the combined system as follows: "I attached sensors to the patient's chest and measured the heart's electrical activity with an electrocardiograph, " or "The patient went into fibrillation, so I placed paddles from a defibrillator and jolted his heart." This is a purely functional description. A purchasing agent at a hospital will describe the same system by listing all the boxes, wires, and prices making up the system. Again, the mechanical and electrical engineers at the manufacturer will describe the system by its mechanism plans and wiring diagrams.

The ability to shift your focus up and down these various levels is crucial to managing the design of a system. A good system design always attempts to group components into discrete levels with as little mixing between

Mathematica program
MathLink template functions
custom functions in C
vendor-supplied library for board
board driver
board physical layer
bits in a chip on the board
electrical signals in sensor
physical phenomenon sensed

Figure 9a   *Mathematica*-linked laboratory system

levels as possible. Properly done, it is possible to assemble components into larger combinations which themselves can be combined – thus allowing even larger systems to be built.

The layering of the laboratory systems described in this book might be something like that in Fig. 9a.

Each layer can have its own local storage and command set. For example, buffers that store acquired data can exist in several locations:

i) in on-board random-access memory. (High-end boards often have several megabytes of on-board RAM so that they can acquire data in high-speed bursts.)

ii) in the driver. (In GUI systems, drivers can execute at "interrupt time," whereas normal applications cannot.)

iii) in a library, for convenience, so that a single function call can deliver a predetermined number of samples.

iv) in a *MathLink* C program, also for convenience, so a single *Mathematica* call can return one or more data sets.

v) in top-level *Mathematica* code.

9.1.2   Polling versus synchronous function calls

Some of the thorniest design problems involve decisions on whether to poll for results or use synchronous hardware control. Polling allows the user to abort or perform other work; synchronous function calls are conceptually easier. Any process taking either significant or unpredictable amounts of time will use one or the other approach (and sometimes both). Improper decisions here will result in either an unresponsive system or a system that is hard to debug. If there is too much synchrony, the program will "hang" for unpredictable amounts of time (possibly forever) and will not feel interactive. If there is too little synchrony, the program's logic will get tangled and unmaintainable. Let us look at some examples to see the problems and opportunities.

9.1.2.1 *Mathematica* code

If you execute the line of *Mathematica* code

In:

```
While[True] Do
```

your copy of *Mathematica* will forever retreat into a world of its own. Nevertheless, you can interrupt the infinite loop by pressing Control-C or Command-period. How does this work?

On UNIX systems, *Mathematica* uses the SIGINT signal to indicate that the user has aborted. On Macintosh systems, there is a "vertical blanking task" that is called every time the monitor screen is refreshed. *Mathematica* installs a custom version of this task to poll the keyboard and see if the user wants to abort.

9.1.2.2 Macintosh Toolbox

In the Macintosh system Device Manager, many of its functions come in two flavors: synchronous and asynchronous. For example, the Macintosh Toolbox that contains system routines has `PBReadSync` – a function that does not return a value until the driver it calls returns. If you use this function, your program will simply wait until it gets an answer before proceeding. Unfortunately, it might never return! Conversely, the function `PBReadAsync` merely starts the reading process; if you use this function, you are responsible for checking whether or not the driver called has finished reading.

9.1.2.3 Drivers

Depending on the design of its associated hardware, a driver can either poll for the state of a memory-mapped hardware flag or issue a command to the hardware that does not return until finished. This is the same problem or opportunity as that found at higher levels of abstraction.

9.1.2.4 Summary

Call synchronous routines (those that do not return until completion) only when their return is guaranteed; that is, there is an associated "status" function that can let you peek and see if the synchronous operation will not block. You can also call synchronous functions if they have a "timeout" parameter, that is, if the function guarantees that it returns within a reliable maximum time. In the latter case you ought to be able to determine from an error code whether something was seriously wrong or whether it was merely a "timeout error," in which case you can try to execute the call again. In several of the examples later in this chapter, a loop is wrapped around a synchronous call that times out in a few seconds. This technique allows the *MathLink* C program to take control and perform useful housekeeping; in particular, it checks to see if the *Mathematica* user wants to abort the function that uses the synchronous call.

9.1.3 Looping

The typical data acquisition task can be described in three steps:

i) set up the hardware,
ii) get the data, and
iii) clean up.

If the data consist of multiple samples, the question arises as to whether "getting the data" involves a single function call that returns multiple samples, a loop that returns a datum for each iteration, or a loop that returns some (but not all) of the data on each iteration. Additionally, if you want your computer to remain responsive to user input – a responsive system – you might want to adopt a polling loop for each of the above cases. You may have several nested loops, with each wrapping one or more synchronous calls. Make sure that you map out a good exit strategy well. The *Mathematica* user may abort at any time, or the hardware might fail. You should also make good use of the message and error mechanisms in *MathLink* to give just the right amount of feedback.

9.1.4 Latency

Related to layers and looping is latency. As a data acquisition program executes, control passes up and down a stack of functions. Data are also passed among these functions. Some of the pathways are fast (a low-level driver "block moves" data from a hardware buffer to the computer's main memory); some are moderately fast (a C program writes to disk or transmits data to *Mathematica* via *MathLink*) and some are extremely slow compared to the hardware acquisition speed (a top-level *Mathematica* function that uses **ListPlot** to graph 20,000 points). You should carefully design your layers so that loops do not cross layers often compared to the fastest loop that needs servicing (usually the hardware clock). You should also avoid latencies resulting from moving large amounts of data through a slow pipe or from extensive processing of data.

9.2 Handling bare hardware (no support libraries)

If you can avoid bare hardware, then do so. Using bare hardware is a miserable job! You will find yourself doing low-level programming and worrying about memory, pointers, timing, synchronization, and other issues. Only marginally better is hardware that has a driver but no high-level libraries to access the driver. In that case, you will not be forced to use assembly language (or assembler-like C), but you will still spend lots of time managing memory and buffers. However, if you make custom hardware or have a special application, you might need to wrap a driver around that hardware – or to just use the truly bare hardware.

In this section, we attack the case of a simple device that has a driver but no support libraries. The device is a simple analog input box for the Apple Macintosh computer. (although the device could equally well be connected to any other computer with a serial port). The device

    i) has two analog input ports and no output,
    ii) is connected to either the modem or printer serial ports,
    iii) is read using the Macintosh serial driver, and
    iv) has absolutely no configuration or flow control.

As soon as the RTS line is asserted (that is, when power is applied and the serial port is opened and reset), the device spews data at a rate of 2400 baud. To make matters worse, it interleaves data from its two ports whether or not both are connected to sensors. That is, it alternately sends to the computer two bytes from channel 0 and two bytes from channel 1. The two high bits of a byte identify the channel; the low six bits of two successive bytes define a 12-bit sample. It is up to the user to synchronize experimental meaning with the data stream; the client program has to hop on to the data stream and figure out where it is.

Nonetheless, this device is successful. It is widely used in the United States in middle- and high-school science classes because it is inexpensive and has a wide range of inexpensive, rugged, and versatile sensors. It has very simple data logging software to capture the data stream. The following example expands upon the serial functions detailed in the previous chapter. As you can see, a simple device can be complicated to manage.

9.2.1 The template file: `SerialDataAcquisition.tm`

First we list the *MathLink* template file. Because it describes the interface between *Mathematica* and the lower-level C code, it acts as a specification for the functionality to be provided, too.

```
:Begin:
:Function:          SerialInitialize
:Pattern:           SerialInitialize[bufferSize_Integer]
:Arguments:         { bufferSize }
:ArgumentTypes:     { Integer }
:ReturnType:        Integer
:End:

:Begin:
:Function:          SerialReadSamples
:Pattern:
SerialReadSamples[samplesRequested_Integer]
:Arguments:         { samplesRequested }
:ArgumentTypes:     { Integer }
:ReturnType:        Manual
:End:

:Begin:
:Function:          SerialPortClose
:Pattern:           SerialPortClose[]
:Arguments:         { }
:ArgumentTypes:     { }
:ReturnType:        Integer
:End:
```

We use a simple design. There is no way to start or stop the serial device, and it is quite slow: 2400 baud is approximately 240 bytes per second which for this device is 60 samples per second per channel. Nonetheless, we do not want to call back and forth across a *MathLink* connection 60 times per second, so we opt to buffer data within the C code. If we allocate a 1MB data buffer (a reasonable size on a modern computer), it will take approximately 4.6 hours to fill up. Therefore, the easiest design is to design a *MathLink* read function that acquires the data synchronously but internally polls data from a lower level. Hence, we have three functions: one to open the port, one to close the port, and one to synchronize to the data stream and acquire a fixed number of samples.

9.2.2   The C code: `SerialDataAcquisition.c`

The narrative for this C code exists both in the C language comments embedded in the code and in text interspersed between code fragments.

```
/*
 * SerialDataAcquisition - this program controls
(barely) a simple low-cost Analog Input device for
Apple Macintosh computers.
 *  It uses the Serial Driver provided with Macintosh
computers to get input.
 *
 * WARNING - This program does NO checking for
"arbitrated" serial ports. If you use AppleTalk
networking, Apple Remote Access, or any of the host of
devices that want to take over the serial port you
will almost certainly crash your machine. */

#include "mathlink.h"
#include <Devices.h>
#include <Serial.h>

#define XONCHAR 0x11
#define XOFFCHAR 0x13

#define OUTDRIVER "\p.BOut"
#define INDRIVER "\p.BIn"

#define BYTESPERSAMPLE 4L

int SerialInitialize (int bufferSize);
OSErr SerialWriteChar (short inVal);
void SerialReadSamples (int samplesRequested);
int SerialPortClose (void);
OSErr SynchronizeDataStream (void);

short gInRefNum, gOutRefNum;
SerStaRec gSerialStatus;
static Ptr gpBuffer = NULL;
static long gBufferSize = 0;
static long gSamplesRead = 0;
```

The above are declarations of functions to match the implied declarations in the `SerialDataAcquisition.tm` template file. Note also the need to have static, global variables that can persist between calls to one of the C functions in this file.

```
int SerialInitialize(int bufferSize)
{
long count;
OSErr err = noErr;

gInRefNum = 0;
gOutRefNum = 0;

if (err = OpenDriver(OUTDRIVER, &gOutRefNum))
    return err;
if (err = OpenDriver(INDRIVER, &gInRefNum))
    return err;
if (err = SerReset(gOutRefNum, baud2400 + data8
                    + stop20 + noParity)) return err;
if (err = SerReset(gInRefNum, baud2400 + data8
                    + stop20 + noParity)) return err;

gBufferSize = bufferSize + 1; // add one so we can
                              //  resync to the data
                              // stream as needed

gpBuffer = NewPtr(gBufferSize * BYTESPERSAMPLE);
if (gpBuffer == NULL) return MemError();
if (err = SerSetBuf(gInRefNum, gpBuffer, gBufferSize *
                    BYTESPERSAMPLE)) return err;

err = SerialWriteChar(255L); // write 'FF' to serial
                             // output to power the
                             // interface
err = err || SynchronizeDataStream();

return (err);
}
```

There is not much complexity there. All of the functions return quickly and are outside the time-critical loop. SerialInitialize simply opens the ports, resets them, and allocates memory for the samples. Note that the memory is assigned to one of the global variables (gpBuffer) declared outside SerialInitialize. This allows us to find the buffer when we subsequently call other C functions in this program.

```
OSErr SerialWriteChar (short inVal)
{
 long num;
 OSErr err = noErr;
 short outbuf;

 num = 1L;
 outbuf = (inVal << 8);

 err = FSWrite(gOutRefNum, &num, &outbuf);
 if (err) return err;
 err = SerStatus(gOutRefNum, &gSerialStatus);
 if (err) return err;

 return (noErr);
}
```

SerialWriteChar is the first foreshadowing of the difficulties with this example. The Macintosh's Device Manager function FSWrite is a synchronous function. We are safe since we have properly reset the serial ports in SerialInitialize and are writing only one datum. If something went wrong, however, we would be stuck within FSWrite with no way to get control!

```
OSErr SynchronizeDataStream(void)
{
 Boolean synchronized = false;
 OSErr err = noErr;
 long count = 0;
 char ch = 0;

 while (!synchronized && (err == noErr))
 {
  err = SerGetBuf(gInRefNum, &count);
  if (count > 0)
  {
   count = 1;
   err = FSRead(gInRefNum, &count, &ch);
   synchronized = (err == noErr) && (ch & 0xC0);
  }
 }
}
```

Because the device we are using continuously reads data and because the only way we can distinguish whether we are reading the high or low bytes of input port 0 or 1 is by their value, SynchronizeDataStream must read bytes until it decides that the next one is the first one we seek. The hardcoded constant 0xC0 identifies the second byte from channel 1. We are going to read exclusively from channel 0, so when we get this value, we know the next byte is where we want to start acquiring data. Note especially that we check the status of the serial port with SerGetBuf. We have to ensure that it is safe to call FSRead – that is, FSRead has received data – and so that our call to it will not hang.

```
void SerialReadSamples(int samplesRequested)
{
 long count = 0;
 long sample = 0;
 short isample;
 int  samplesProcessed = 0;
 OSErr err = noErr;

 MLPutFunction(stdlink, "List", samplesRequested);
 while ((MLAbort == 0) &&
         (samplesProcessed < samplesRequested) &&
         (err == noErr))
 {
  err = SerGetBuf(gInRefNum, &count);
  if (count >= 4)
  { // something to process!
   count = 4;
   err = FSRead(gInRefNum, &count, &sample);
   sample = sample >> 16; // wipe out
                          // channel 1 (Port 2)
   isample = (short)(((sample & 0x3F00) >> 2) |
                     // high 6 bits
                     (sample & 0x003F)); // low 6 bits
   MLPutInteger(stdlink, isample);
   samplesProcessed++;
  }
 }
 if (MLAbort == 1) MLPutFunction(stdlink, "Abort", 0);
 MLEndPacket(stdlink);
}
```

SerialReadSamples begins a *Mathematica* **List** and then repeatedly loops. In the body of the loop it peeks at the serial port's buffer to see if it can read without hanging. If so, it reads 4 bytes into a single long integer.

Remember that we are synchronized so we can tell which bits are our desired sample. We throw away the two bytes corresponding to the channel we are not monitoring and then extract the bits we want, combining them into a single 12-bit integer. We return this integer to *Mathematica* with `MLPutInteger`. Note that the repeated calls to this *MathLink* function yield sufficient time to the *MathLink* library for it to propagate abort messages. If an abort message is generated by the user, then `MLAbort` will become equal to 1 and we can exit the loop and call **Abort[]**.

```
int SerialPortClose(void)
{
 OSErr err = noErr;
 OSErr killErr;

 killErr = KillIO(gInRefNum);
 killErr = KillIO(gOutRefNum);

 if (gpBuffer != NULL)
 {
  DisposePtr(gpBuffer);
  gpBuffer = NULL;
 }

 err = SerSetBuf(gInRefNum, gpBuffer, 0);
 if (gInRefNum)
  err = err | CloseDriver(gInRefNum);
 if (gOutRefNum)
  err = err | CloseDriver(gOutRefNum);
 return (err);
}

#if !WINDOWS_MATHLINK

int main(argc, argv)
        int argc; char* argv[];
{
        return MLMain(argc, argv);
}

#else

int PASCAL WinMain( HANDLE hinstCurrent, HANDLE
hinstPrevious, LPSTR lpszCmdLine, int nCmdShow)
```

```
{
 char  buff[512];
 char FAR * argv[32];
 int argc;

 if( !MLInitializeIcon( hinstCurrent, nCmdShow))
    return 1;
 argc = MLStringToArgv( lpszCmdLine, buff,
    argv, 32); return MLMain( argc, argv);
}
#endif
```

These routines are merely housekeeping. In later examples, we will intentionally make sure that we clean up automatically when an error or abort condition occurs.

9.2.3   Using SerialDataAcquisition

We begin by allocating a buffer big enough for 5000 samples. The zero returned value indicates that no error occurred.

In:

SerialInitialize[5000]

Out:

    0

Next, we take 250 samples (about 5 seconds' worth) from a slow-reacting thermometer. We wrap the call to **SerialReadSamples** with a call to **ListPlot**; the resulting plot is not interesting – but you can see that the temperature is stable.

In:

SerialReadSamples[250]

Out:

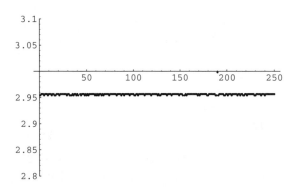

Lastly, we close the serial port.

In:

```
SerialPortClose[]
```

Out:

```
0
```

In the next section, we explore the more common case of data acquisition and control mediated by vendor-supplied libraries. Because this is the case that engineers, scientists, and students encounter more commonly, we will build a library of useful routines and techniques to help you make polished *MathLink* template programs.

9.3   Handling hardware with vendor libraries

Most vendors include a library of common data-acquisition and control functions with each board sold, if for no other reason than to provide the user with diagnostic routines. Traditionally, such libraries are meant to be called from FORTRAN, C, Pascal, or BASIC. As you may expect, given a suitable C library we can hook such a board to *Mathematica* via *MathLink*.

In this section, we examine several example using the NI-DAQ libraries from National Instruments. NI-DAQ (which stands for National Instruments Data AcQuisition) is available for a wide range of data acquisition hardware for DOS, Windows, and Macintosh systems. It provides functions in many application areas including:

- analog input
- analog output
- digital input and output
- data acquisition (single- and double-buffered)
- counter/timer functions
- waveform generation.

NI-DAQ handles many boards of differing capabilities with a common programming interface. For the examples given below, we will use a LAB-LC board installed in a Power Macintosh 5215 computer. Since the NI-DAQ calls used are cross-platform for Windows and Macintosh, and since the *MathLink* C code used is also cross-platform, the code samples in this chapter should work unchanged with other National Instruments boards on Macintosh or Windows machines.

9.3.1   One-shot oscilloscope

A one-shot oscilloscope acquires data in what is known as "burst mode". That is, it acquires a predetermined amount of data in response to some event. This event (called the "trigger") may simply be the acquisition command itself (known as a software trigger) or an externally generated trigger event. External triggers are used when one wants only to take measurements in response to some experimental or physical condition; for example, a laser pulse might be used to trigger observations of the fluorescence of a gas. In this case, a sensor attuned to the laser pulse generator would start acquisition on a sensor monitoring the target gas. Triggering restricts the data acquired to a time period near the time of interest. In NI-DAQ, there are many ways to specify software or external triggering. For example, the function DAQ_Config takes as its second parameter 0 or 1 to indicate a software or external trigger, respectively. In the following example, we hardcode a software trigger in our *MathLink* C code. It is a simple matter to expose this to *Mathematica* by adding a parameter in the *MathLink* template file. Let us now look at the files that comprise a one-shot oscilloscope and see some examples of its use.

9.3.1.1   The template file: OneShotScope.tm

```
:Begin:
:Function: oneShotScope
:Pattern:   OneShotScope[slotNum_Integer,
               channel_Integer, sampleInterval_Integer,
               sampleCount_Integer]
:Arguments: {slotNum, channel, sampleInterval,
               sampleCount}
:ArgumentTypes:   { Integer, Integer, Integer, Integer }
:ReturnType: Manual
:End:
```

9.3.1.2   The C code: OneShotScope.c

This template is about as simple as possible. The slotNum argument corresponds to the backplane slot in the computer in which the LAB-LC board resides. Because most computers can support multiple installed boards, it is usually necessary to indicate which board should receive the command. The channel argument corresponds to the analog input channel on the board; most boards support multiple channels of data acquisition. The sampleInterval argument represents the time between measurements. For the National Instruments line of boards, sampleInterval is measured in units of the base, which may be set over different ranges. For this example, base is set to 1 μs. Finally, sampleCount is simply the number of data points desired. Note that the ReturnType is Manual since we will return a list of input voltages.

```
/*
 * OneShotScope.c  -  This example performs a single
channel data acquisition operation using NI_DAQ
interface  routines.
 */

#include <stdio.h>
#include <memory.h>
#include "NI_DAQ_MAC.h"
#include "mathlink.h"
#include "MitlUtil.h"

void oneShotScope(int16 board, int16 channel, int32
sampleInterval, int32 sampleCount);

/*......acquisition parameters........*/
int16
   gain = 1,    /* use gain = 1 */
timebase = 1;/* timebase = 1 ==> 1 μs timebase */

void oneShotScope(int16 board, int16 channel,
           int32 sampleInterval, int32 sampleCount)
{
 int16
   i,
error,
status,  /* used in call to DAQ_Check to reflect
           acquisition completion status */
*buffer;  /* buffer for the acquired, unscaled data */

 float *volt_array; /* buffer for the scaled data */
 int mlerr;

MITL_TRY
/*............begin.................*/
/* NULL the buffer pointers */
 buffer = NULL;
 volt_array = NULL;

 /* allocate the buffer for the acquired data and the
```

```
buffer for the scaled data */
  buffer = svector(sampleCount); /* allocate
                                      integers */
  volt_array = vector(sampleCount); /* allocate
                                        floats */

 /* clear any current acquisition */
  error = DAQ2Clear(board);


 /* start the acquisition */
  error = DAQ_Start(board, channel, gain, buffer,
             sampleCount, timebase, sampleInterval);

  chkerr("DAQ_Start", error);
  if (error)
  {
   error = DAQ_Clear(board);
   chkerr("DAQ_Clear", error);
  }

 /* wait for the acquisition to finish */
  status = 0;
  while ((status == 0) && (error == 0) &&
          (MLAbort == 0))
  {
   /* make sure we can get the MLAbort flag set */
   MLCallYieldFunction(MLYieldFunction(stdlink),
                       stdlink,
                       (MLYieldParameters)0);
   /* check to see if we got the data or an error */
   error = DAQ_Check(board, &status);
   chkerr("DAQ_Check", error);
  }

  if (MLAbort != 0)
   MLPutFunction(stdlink, "Abort", 0);
  else
  {
  /* scale the data */
   error =  DAQ_Scale(board, gain, sampleCount, buffer,
                      volt_array);
   chkerr("DAQ_Scale", error);

  // send the floating-point data
```

```
    {
     int32 i;
     double tempDouble;
     float *pFloat = volt_array;
     MLPutFunction(stdlink, "List", sampleCount);
     for (i = 0; i < sampleCount; i++)
     {
      tempDouble = *pFloat++;
      MLPutFloat(stdlink, tempDouble);
     }
     MLEndPacket(stdlink);
    }
   }

   free_svector (buffer);
   free_vector (volt_array);
  MITL_RECOVER
   MLPutSymbol(stdlink, "$Failed");
  MITL_ENDTRY
  return;

 } /* oneShotScope */
```

The function `oneShotScope` allocates a vector of short integers to hold the raw data from the board. It also allocates a vector of floating-point numbers to hold the converted data. Conversion is done by the NI-DAQ libraries. If you use a library without such conversion, you can convert the numbers according to your hardware's specifications with C code, or else you can send the raw data to *Mathematica* and convert it there. After allocating these two vectors (which are discussed further later), note both the call to `DAQ_Start` to start the acquisition and the loop that repeatedly calls `DAQ_Check` to see if the data have arrived. Upon successful completion of the acquisition, the raw data are submitted to `DAQ_Scale` for conversion to floating-point numbers. We use this call because the driver that controls the hardware also contains information about the gain of the hardware; that is, the driver knows the mapping between the integer data and the physical floating-point voltages. Next, we send the floating-point data to *Mathematica* by looping over `MLPut-Float`. Note that we cannot use `MLPutRealList` since *Mathematica* considers **Real** to be a double-precision floating-point number (`double`) rather than a single-precision number (`float`). Finally, we free up the buffers.

At the end of this chapter is the source listing for `MitlUtil.h` and `MitlUtil.c`, which define the macros `MITL_TRY`, `MITL_RECOVER`, and `MITL_ENDTRY` as well as the memory allocation routines `svector`, `vector`, and so on. These routines will recover gracefully from memory allocation (or other) errors and will propagate error messages back to *Mathematica*.

9.3.1.3   Using OneShotScope

Now let us see what happens when we invoke OneShotScope within *Mathematica*. Install the template program as always:

In:

```
1 = Install["OneShotScope"]
```

Out:

```
LinkObject[OneShotScope, 6, 2]
```

Note that the function OneShotScope now exists.

In:

```
LinkPatterns[1]
```

Out:

```
{OneShotScope[slotNum_Integer, channel_Integer,
sampleInterval_Integer, sampleCount_Integer]}
```

Let us acquire data from channel 4 on board 6 with a time interval of 25 times the base of 1 μs (which corresponds to 40 kHz). We acquire 250 samples. The result is fed directly into **ListPlot**; the data could just as easily be assigned to a *Mathematica* symbol for further processing.

In:

```
ListPlot[OneShotScope[6,4,25,250],
        PlotJoined->True]
```

Out:

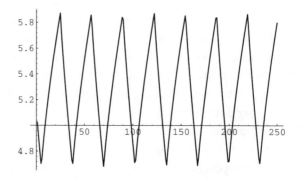

As you may have guessed, channel 4 is hooked up to a sawtooth wave source. Similarly, we can acquire data on another input channel. (In this case, channel 3 is connected to a square wave source.)

In:

```
ListPlot[OneShotScope[6,3,25,250],
        PlotJoined->True]
```

Out:

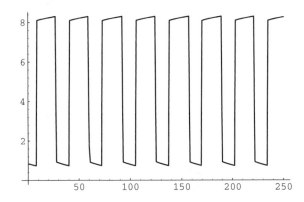

Finally, we see how gracefully we handle a request for 1 million samples (which is far more than the external *MathLink* program can handle, since it was arbitrarily set to have a maximum memory partition of 500 kbytes). Notice that the program generates *Mathematica* error messages and recovers.

In:

```
ListPlot[OneShotScope[6,3,25,1000000]]
```

Out:

```
MitL::mitlerr:
    MitL run-time error; allocation failure in svector().
ListPlot::list:
    List expected at position 1 in ListPlot[$Failed].
ListPlot[$Failed]
```

We can now uninstall the program normally (it did not die when we attempted the huge acquisition above).

In:

```
Uninstall[1]
```

Out:

```
OneShotScope
```

9.3.1.4 Moving on

You might want to extend `OneShotScope` to handle multiple channels simultaneously. You will need to allocate an `svector` and a `vector` for each channel and to use `Lab_SCAN_Start` to start a multi-channel acquisition.

Use Lab_SCAN_Check to check on the status of the acquisition. When the data are acquired, you can either use SCAN_Demux to demultiplex the data into separate buffers or send the whole mess back to *Mathematica* and use **Partition** and **Transpose** to split up the data. Warning: if the channels use different scales to map between integers and floating-point voltages, you will need to demultiplex and scale the data in the *MathLink* C program before returning the data to *Mathematica*.

9.3.2   Continuous acquisition (one channel)

Many laboratory applications require continuous acquisition of data over an extended period of time, where "extended" means "more data than can fit in the computer's memory at once" or "longer than I'm willing to wait without feedback." For example, a medical laboratory might need to monitor the temperature of its blood bank over many days. These applications usually need to update a strip chart or other output device so the user can periodically monitor the process without disturbing the acquisition.

Continuous acquisition is a challenge for *Mathematica* since *Mathematica* is inherently a batch-style system. That is, *Mathematica* is driven by a read – evaluate – print loop: one question yields one answer, although that answer might be a compound answer with many parts (such as the **List** of numbers returned by **OneShotScope**). Note that *Mathematica* plots are not like a strip chart. They spring completed from *Mathematica*. How then are we to get continuous feedback, mimicking a strip chart application? Apropos to our earlier discussion about loops and layers, the logical approach is to write our loop in top-level *Mathematica* code. Each pass through the loop will acquire more data through a *MathLink* C program and plot it. There are, however, two difficulties that we must overcome. First, how do we keep on the data acquisition hardware as we go back and forth between *Mathematica* and the *MathLink* template program? Second, how do we avoid swamping *Mathematica* with megabytes of data during the long acquisition?

To handle the first problem, we need to implement three functions in our *MathLink* template program: one to start acquisition, one to stop acquisition, and one to get the next buffer of data from the running hardware. The second problem can be handled either by throwing the data away after plotting it or by routing it to a disk file for later analysis. In this example we will take the former course; in a later example we will see how to retain data in a disk file.

Starting and stopping continuous data acquisition hardware is quite simple; virtually all libraries, NI-DAQ included, provide functions to do just that. In fact, it is quite difficult to see how one could have a useful product without such functions! The conceptually tricky part is getting the data. Once the board is started, the data flow at a continuous rate; however, we need discrete "buckets" of data. We also need to ensure that we are not emptying a bucket that the hardware is still trying to fill. How can we accomplish this? The solution is straightforward: by having two or more buckets, rather like a bucket brigade carrying water to a fire. Full buckets travel one at a time from the well, are emptied, and return to the well to be filled again. Similarly, we can provide two memory buffers. The hardware fills one, hands it off to the user's program, and proceeds to fill the other; meanwhile, the *MathLink* template program copies the full buffer, transmits the data to *Mathematica*, and yields the buffer back to the hardware. Such an approach is called "double-buffered acquisition."

In the example below, we will use NI-DAQ routines to manage a double-buffered acquisition; NI-DAQ can create both buffers in its driver's memory space; this saves us the necessity of maintaining and managing a ring buffer and pointer indices. We then need only to tap into the buffers when they are ready. If your vendor-supplied library does not provide you with such assistance, you will need to create two or more memory buffers and swap them yourself. You will also need to check for overrun errors resulting from the software process not keeping up with the hardware.

We now turn to the actual *MathLink* template and C code, which is only slightly more complex than the first example.

9.3.2.1 The template file: `Oscilloscope.tm`

```
:Begin:
:Function: startAcquisition
:Pattern: StartAcquisition[slotNum_Integer,
                           channel_Integer,
                           sampleInterval_Integer,
                           sampleCount_Integer,
                           blockSize_Integer]
:Arguments: { slotNum, channel, sampleInterval,
              sampleCount, blockSize }
:ArgumentTypes:  { Integer, Integer, Integer, Integer,
                   Integer }
:ReturnType: Manual
:End:

:Begin:
:Function: getNextBuffer
:Pattern: GetNextBuffer[slotNum_Integer,
                        channel_Integer,
                        sampleCount_Integer]
:Arguments: { slotNum, channel, sampleCount }
```

```
:ArgumentTypes: { Integer, Integer, Integer }
:ReturnType: Manual
:End:


:Begin:
:Function: stopAcquisition
:Pattern: StopAcquisition[slotNum_Integer]
:Arguments: { slotNum }
:ArgumentTypes: { Integer }
:ReturnType: Manual
:End:
```

As you can see, three functions are implemented in this template – one to start, one to stop, and one to get a data buffer. Each function takes a slot number to identify the board, although that could easily be hardcoded in the *MathLink* C code. **StartAcquisition** and **StopAcquisition** return the symbol **Null** if everything works correctly. **GetNextBuffer** returns a list of numbers. A more robust implementation of this example would have **StartAcquisition** return an **AcquisitionLink[]** object that is used as the first parameter to **GetNextBuffer** and **StopAcquisition**. This approach, which would hide more of the details and would parallel the design of *MathLink* itself, is left as an exercise.

9.3.2.2 The C code: Oscilloscope.c

```
/*
 * Oscilloscope.c  -  Performs double-buffered
 continuous acquisition
 */

#include <stdio.h>
#include <memory.h>
#include "NI_DAQ_MAC.h"
#include "mathlink.h"
#include "MitlUtil.h"

#define SLOPE     0          /* TRIGGER - trigger on
                                a negative slope */
#define VALUE     0          /* when the value
                                0 is passed */
#define TRIGPOS   50L         /* at the 50th position
                                in the buffer */

void startAcquisition(int16 board, int16 channel, int32
```

```
                          sampleInterval, int32 bufferSize,
                          int32 blockSize);
void getNextBuffer(int16 board, int16 channel,
                   int32 blockSize);
void stopAcquisition(int16 board);

static int32COUNT = 0;/* ignored for continuous
                         acquisition - so set to 0 */
static int16gain = 1;
static int16timebase = 3;/* use a 10KHz clock */
static int16cntmode = 1; /* enable continuous data
                            acquisition */
static int16*buffer = NULL;
static float*volt_array = NULL;

void startAcquisition(int16 board, int16 channel,
                      int32  sampleInterval,
                      int32 bufferSize, int32 blockSize)
{
int16 error;

MITL_TRY

/* clear any current acquisition */
error = DAQ2Clear(board);

/* acquire memory for acquisition buffers */
if (buffer != NULL) free_svector(buffer);
if (volt_array != NULL) free_vector(volt_array);
buffer = svector(blockSize);
volt_array = vector(blockSize);

/* configure for continuous block mode data acquisition
*/
error = DAQ2Config(board,cntmode,bufferSize, blockSize);
chkerr("DAQ2Config", error);

/* start a single channel data acquisition */
/* NOTE: since data are acquired continuously, pass
in 0 for buffer and count parameters */
error = DAQ_Start(board, channel, gain, NULL, COUNT,
                  timebase, sampleInterval);
chkerr("DAQ_Start", error);
MLPutSymbol(stdlink, "Null");
MITL_RECOVER
```

```
          MLPutSymbol(stdlink, "$Failed");
          MITL_ENDTRY

          return;
          }

          void getNextBuffer(int16 board, int16 channel,
                          int32 blockSize)
          {
          int16error;
          int32actualCount; /* used by DAQ2TGet */

          MITL_TRY
            /* acquire the most recent block of data to graph */
            error = timeOutErr;/* assume that the data isn't
                              ready yet */
            while ((MLAbort == 0) && (error == timeOutErr))
            {
              MLCallYieldFunction(MLYieldFunction(stdlink),
                              stdlink, (MLYieldParameters)0);
          /* try to get the data - make sure we timeout every 5
          seconds so we can pop out of DAQ2TGet and check the
          abort status */
              error =DAQ2TGet(board, channel, SLOPE, VALUE,
                          buffer, TRIGPOS, blockSize,
                          &actualCount, 300L);
            }
            if (MLAbort != 0)
              MLPutFunction(stdlink, "Abort", 0);
            else
            {
              chkerr("DAQ2TGet",error);
              if (error == 0)
              {
                /* scale digital values to voltages */
                error = DAQ_Scale(board, gain, blockSize, buffer,
                              volt_array);
                chkerr("DAQ_Scale", error);
                /* send the floating-point data */
                if (error == 0)
                {
                  int32 i;
                  double tempDouble;
                  float *pFloat = volt_array;
                  MLPutFunction(stdlink, "List", blockSize);
```

```
                for (i = 0; i < blockSize; i++)
                {
                  tempDouble = *pFloat++;
                  MLPutFloat(stdlink, tempDouble);
                }
                MLEndPacket(stdlink);
              }
              else
                MLPutSymbol(stdlink, "$Failed");
          }
          else
            MLPutSymbol(stdlink, "$Failed");
      }
      MITL_RECOVER
      MLPutSymbol(stdlink, "$Failed");
      MITL_ENDTRY

  return;
  }
  void stopAcquisition(int16 board)
  {
    int16error;

    error = DAQ2Clear(board);
    chkerr("DAQ2Clear", error);

    free_svector(buffer);buffer = NULL;
    free_vector(volt_array);volt_array = NULL;

    MLPutSymbol(stdlink, "Null");

    return;
  }
```

The three functions in this example are quite simple. First, startAcquisition allocates an integer and a floating-point buffer of our desired block (bucket) size. It then configures the hardware for doublebuffered acquisition using DAQ2_Config and starts the acquisition with DAQ_Start. Note that DAQ2_Config also takes a bufferSize parameter. The NI-DAQ library uses bufferSize to allocate the "bucket brigade" alluded to before. We are responsible only for allocating a single bucket, which we will use to "tap" into the brigade. For example, if bufferSize is 10000 and blockSize is 500, then the NI-DAQ library will allocate room for 10000 samples and deliver in 500-sample blocks. The buffer that NI-DAQ allocates is a circular, or ring, buffer. When the hardware gets to the end of the buffer, it goes back to the beginning and continues filling the buffer. The NI-DAQ library and driver maintain pointers into this ring buffer; these pointers remember where the hardware is writing and where the user is reading. If one does not get the data out of the buffer fast enough, the data will be overwritten and lost. On the other hand, overwriting may or may not matter, depending on the application. In particular, if the data lost represents a time period that is small compared to the events of interest, we may not concern ourselves with the lost data. Also note in passing that the hardcoded timebase is different than the value used in **OneShotScope**. Again, if you want flexibility, the timebase should be an additional parameter in the *MathLink* template.

In `getNextBuffer`, the C code uses the NI-DAQ routine `DAQ2TGet`, which gets the next block in the ring buffer created and maintained by the driver. If the hardware is acquiring data faster than *Mathematica* can process them, the board will start to overwrite blocks before you can retrieve them – an overrun error. If you do not need to get all the data and need to keep up only with the most recent block, then the NI-DAQ function `DAQ2TTap` can be used to retrieve the most recently filled block. By using `DAQ2TTap`, you will get gaps in the acquired data stream. If we sample a periodic source that is much faster than *Mathematica*'s processing speed, an animation of the data will jitter back and forth much like a real oscilloscope connected to, but failing to trigger from, a source. The only other item to note in `get-NextBuffer` is the special treatment given to the error code `timeOutErr`. This error, indicating that the block is not ready, occurs when *Mathematica* is faster than the hardware; in that case, we simply have to try again. The user of **GetNextBuffer** must be prepared to handle this condition. The example puts a loop inside the C function `getNextBuffer` to repeatedly try `DAQ2TGet`. It also checks `MLAbort` so that it can interrupt slow acquisitions! See Chapter 7, Section 7.11.2 if you need to remember how to make *MathLink* C functions interruptible.

9.3.2.3 Using `Oscilloscope`

Here are some examples of continuous acquisition. First, we get nine buffers of data from an continuously acquired 1250-Hz sawtooth-wave. The call to **StartAcquisition** reads as follows: "Start to acquire data from the board in slot 6 with a sample interval of 8 clock ticks (an $800\,\mu s$ period implies a frequency of 1250 Hz), allocate a ring buffer for 10000 samples, and use a block size of 250 samples."

In:

```
StartAcquisition[6,4,8,10000,250]
```

```
sawData = Table[GetNextBuffer[6,4,250],{9}];
StopAcquisition[6]
```

We now **Map** the function **ListPlot** over the returned data to get nine plots. If you run the resulting animation you will find that it looks just like an oscilloscope display.

In the interests of space, we show only the first plot full-size followed by a **GraphicsArray** of all the data. Note that the plots in the **GraphicsArray** are in fact continuous. We have not lost any data.

In:

```
sawPlots = Map[ListPlot[#, PlotRange->{4.6,5.9}]&,
sawData]
```

Out:

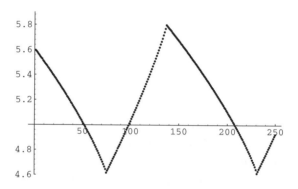

In:

```
Show[GraphicsArray[Partition[sawPlots,3]]]
```

Out:

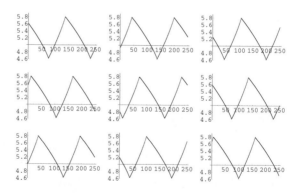

Next, we use a slow acquisition taken from a thermometer at room temperature over the span of 500 seconds. As you can see, not much is happening! Even a sampling rate of 2 Hz is too fast for this application.

In:

```
StartAcquisition[6,1,5000,10000,250]
tempData = Table[GetNextBuffer[6,1,250],{5}];
StopAcquisition[6]
tempGraphs = Map[ListPlot[#,
                          PlotRange->{2.8,3.1}]&,
              tempData]
```

Out:

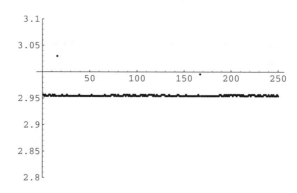

Finally, we show the case of an aborted acquisition. Note that we abort only the call to **GetNextBuffer**. When selecting and executing all three *Mathematica* statements together, it is nice to have the **StopAcquisition** function execute even if we abort **GetNextBuffer**.

In:

```
StartAcquisition[6,1,5000,10000,250]
tempData = Table[GetNextBuffer[6,1,250],{5}];
```

Out:

```
$Aborted
```

In:

```
StopAcquisition[6]
```

9.3.3 Streaming data to disk

In the previous example, we used continuous double-buffered acquisition to implement a simple logging process. The specific implementation, however, could drop data since it used DAQTTap to get the most recently acquired block of data from the driver's ring buffer. Moreover, even if we had used DAQ2Get, we might not be able to keep up with the hardware since *MathLink* and *Mathematica* are much slower than data acquisition hardware and a tuned driver written in C or assembler.

In this example, we use double-buffered acquisition to stream data directly from the hardware to a disk file. Hence the only information that travels over a *MathLink* connection to *Mathematica* is control and status information, which can be orders of magnitude less voluminous than the actual data stream. Since we have laid the groundwork for this example in the previous one, we can dive into the code; the only novel twist is managing the disk files. This example uses portable file-handling functions from the ANSI C library; such functions are declared in `stdio.h`. Applications that have special requirements can also use the platform-native file-handling routines or custom file-handling libraries.

9.3.3.1 The template file: `StreamToDisk.tm`

```
:Begin:
:Function:        streamToDisk
:Pattern:         StreamToDisk[filename_String,
                               slotNum_Integer,
                               channel_Integer,
                               sampleInterval_Integer,
                               sampleCount_Integer]
:Arguments:       { filename, slotNum, channel,
                    sampleInterval, sampleCount }
:ArgumentTypes:   { String, Integer, Integer, Integer,
                    Integer }
:ReturnType:      Manual
:End:
```

This template is identical to **OneShotScope** with the addition of a parameter denoting the file into which we wish to stream the data. Note that **StreamToDisk** is like **OneShotScope** in that it acquires a fixed number of samples; it is also like **StartAcquisition, GetNextBuffer**, and **StopAcquisition** in that it uses double-buffered acquisition to handle more data than can be acquired in a single buffer. In passing, we should note that you could stream data to disk by writing a small *Mathematica* program that repeatedly calls **GetNextBuffer** and the **WriteBinary** function from the standard *Mathematica* package `Utilities`BinaryFiles`. This example basically handles getting the data and writing the data to disk within a single *MathLink* function. As always, the trade-off is between performance and flexibility.

9.3.3.2 The C code: `StreamToDisk.c`

```
/*
 * StreamToDisk.c
```

```
  *
  *This routine is an example of a single channel data
acquisition using the double
  *buffered method of acquisition (DAQ2).
  *As each block of data is acquired (using DAQ2Get)
it is written to disk.
  *
  */
#include <stdio.h>
#include "NI_DAQ_MAC.h"
#include "mathlink.h"
#include "MitlUtil.h"

/* too handy not to have! */
#if defined(false)
  #if false != 0
    #error Incompatible 'false' already #defined
  #endif
#else
  #define false 0
#endif

#if defined(true)
  #if true != 1
    #error Incompatible 'true' already #defined
  #endif
#else
  #define true 1
#endif

typedef unsigned char Boolean;

/*................these are the fixed parameters
.....................*/
int16cntMode = 0,/* disable continuous acquisition */
timebase = 3,/* 100 μsec resolution */
gain = 1;/* gain of 1 */

int32blkBufSize = 5000L,/* size of block buffer in
number of samples (2 bytes each) */
circBufSize = 100000L;/* size of circular buffer in
number of samples (2 bytes each) */

/* called to free up memory and close files */
void CleanUP (int16 board, int16 *blkBuf, FILE *fileRef,
```

```
Boolean failed, char *filename);
void streamToDisk (char *filename, int16 board, int32
channel, int32 sampleInterval, int32 sampleCount);

void streamToDisk (char *filename, int16 board,
                   int32 channel, int32 sampleInterval,
                   int32 sampleCount)
{
  FILE*fileRef = NULL;
  int16err = noErr;
  int16*blkBuf = NULL;   /* pointer to buffer DAQ2Get
                            writes data to */
  int32 actualCount;     /* used by DAQ2TGet */

  int32 samples_to_get;/* number of samples in a block
                          returned by DAQ2Get */
  int32samples_remaining;/* number of total samples
                           left to acquire */
  int32 samples_written; /* numbers of samples actually
                            written by a call to
                            fwrite*/
  MITL_TRY
/* open the data file on disk */
  fileRef = fopen (filename, "wb");

  if (fileRef == NULL)
  {
    mitlerror("Error opening file in fopen.");
    MLPutSymbol(stdlink, "$Failed");
    return;
  }

  blkBuf = NULL;
  blkBuf = svector(blkBufSize);

  DAQ2Clear (board);/* always a good idea to call a
                       Clear function before any of the
                       other DAQ functions */

/* configure for double buffered mode */
  err = DAQ2Config (board, cntMode, circBufSize,
                    blkBufSize);
  chkerr ("DAQ2Config",err);
```

```
/* initialize the samples remaining value and some
file io parameters*/
  samples_remaining = sampleCount;

  err = DAQ_Start(board, channel, gain, NULL,
                  sampleCount, timebase,
                  sampleInterval);
  chkerr( "DAQ_Start",err);
  if (err == 0)
  {
/* loop until done, getting blocks of data and writing
to disk */
  while (samples_remaining)
  {
  samples_to_get = (samples_remaining < blkBufSize) ?
                     samples_remaining : blkBufSize;
  samples_remaining -= samples_to_get;

/* all the parameters concerned with triggering are set
to 0 */
  err = timeOutErr;/* assume data aren't ready yet */
  while ((MLAbort == 0) && (err == timeOutErr))
  {
    MLCallYieldFunction(MLYieldFunction(stdlink),
                        stdlink, (MLYieldParameters)0);
    err = DAQ2TGet (board, channel, 0, 0, blkBuf, 0,
    samples_to_get, &actualCount, 300L); /* 5 second
                                           timeout */
  }
  if (err)
  {
    chkerr ( "DAQ2TGet",err);
    throw_on_error("Error in DAQ2TGet.");
  }

  samples_written = fwrite (blkBuf, sizeof(int16),
                           samples_to_get, fileRef);
  if (samples_written != samples_to_get)
  {
    throw_on_error("Error from fwrite.");
  }
  } /* get block and disk storage while loop */
}
```

```
/* normal clean-up */
  CleanUP (board, blkBuf, fileRef, false, filename);
  return;
  MITL_RECOVER
  CleanUP (board, blkBuf, fileRef, true, filename);
  MITL_ENDTRY

} /* DAQ2Example */

void CleanUP (int16 board, int16 *blkBuf, FILE *fileRef,
            Boolean failed, char *filename)
{
/* clear the board configuration table */
  DAQ2Clear (board);

/* free the buffer space */
  if (blkBuf != NULL)
  {
    free_svector (blkBuf);
    blkBuf = NULL;
  }

/* close the data file and flush volume info */
  fflush (fileRef);
  fclose (fileRef);

/* send the appropriate answer to Mathematica */
  if (failed)
    MLPutSymbol(stdlink, "$Failed");
  else
    MLPutString(stdlink, filename);

  return;
}
```

Again, this is a very simple program. After allocating buffers, opening a disk file, and initializing the board, the C code repeatedly gets a block of samples using DAQ2Get and writes the block to disk using the standard C library function fwrite. Note that we ensure some responsiveness by using the timeout feature of DAQ2Get; by passing a nonzero value for timeout (the last parameter), we ensure that we periodically pop out of DAQ2Get. We can check to see if the error was simply a timeout error, and we also check to see if the user aborted. Additionally, we could send a packet to *Mathematica* to print some progress information; this is omitted here for clarity. The important gain is that we have preserved *Mathematica*'s interruptibility. Had we used zero for the timeout, we would have made it impossible to gracefully interrupt the acquisition.

The other programming point to note is the function `CleanUP`. This function gathers together memory cleanup and file flushing and closing. It additionally returns either **$Failed** or a string denoting the name of the file into which the data is saved. Therefore, the result of **StreamToDisk** can be used in the function **OpenBinaryFile** from the standard *Mathematica* package `Utilities 'BinaryFiles'`. One can also use the **FastBinaryFiles** *MathLink* package on MathSource.

9.3.3.3 Using `StreamToDisk`

StreamToDisk takes the same parameters as **OneShotScope**, plus a filename. The example below asks **StreamToDisk** to acquire 250,000 samples from channel 4 of the board in slot 6. The sample interval is $8 * 100\,\mu s$. This acquisition will take 200 seconds. Note that the file size is in fact 500,000 bytes, which is correct because samples are two bytes per sample.

In:
```
theFile = StreamToDisk["lotsof.dat",6,4,8,250000]
```
Out:
```
lotsof.dat
```
In:
```
FileByteCount[theFile]
```
Out:
```
500000
```

9.3.3.4 Moving on

You can modify **StreamToDisk** to save 4-byte floating-point voltages. It currently saves the raw 2-byte integers. You will need to use `DAQ_Scale` to convert the voltages. Remember to create a **vector** as well as an **svector** and to modify the call to `fwrite`.

You might also consider extending **StreamToDisk** by incorporating code from the *MathLink* package `MathHDF`, which supports reading and writing to HDF files. HDF is a portable binary file format that supports large scientific data sets. It is fast and efficient and is used widely in the scientific programming community. A version of `MathHDF` can be found on the MathSource web site (`http://www.wri.com`) as item number 0203-555. Since HDF is a well-supported file format and library with extensive ongoing development, interested readers should also check the HDF home page (`http://hdf.ncsa.uiuc.edu`) to obtain the latest versions of the library.

StreamToDisk can also be extended so that it streams data from multiple channels into multiple files simultaneously. You can use `DAQ2Config`, `SCAN_Start`, and `SCAN_Setup` to set multiple-channel acquisition. Get the data with `DAQ2TGet` and then demultiplex (separate) the data into multiple buffers with `SCAN_Demux` before writing the demultiplexed data buffers into the files.

9.3.4 Generating a waveform

Previous examples have focused on data acquisition and analysis in *Mathematica*. Now we show how to accomplish the opposite. That is, we show how to output a waveform generated in *Mathematica*. Instead of generating square and sine waves you can use *Mathematica* to generate and output Airy, Bessel, or Legendre functions. Since *Mathematica* can accurately generate data, it is a simple matter to send the data to an external *MathLink* program which will load the data into the hardware and ask it to repeatedly convert the binary digits into analog voltages. By the way, for this example we will massage the waveform in *Mathematica*; this shows how easy it is to do routine data transformations in *Mathematica*. First, here are the template and C files.

9.3.4.1 The template file: `GenerateWaveform.tm`

```
:Begin:
:Function:          setupWaveform
:Pattern:           SetupWaveform[slotNum_Integer,
                                  channel_Integer,
                                  sampleInterval_Integer,
                                  waveform_List]
:Arguments:         { slotNum, channel, sampleInterval,
                      waveform }
:ArgumentTypes:     { Integer, Integer, Integer,
                      IntegerList }
:ReturnType:        Manual
:End:

:Begin:
:Function:          startWaveform
:Pattern:           StartWaveform[slotNum_Integer]
:Arguments:         { slotNum }
:ArgumentTypes:     { Integer }
```

```
:ReturnType:        Manual
:End:


:Begin:
:Function:          stopWaveform
:Pattern:           StopWaveform[slotNum_Integer]
:Arguments:         { slotNum }
:ArgumentTypes:     { Integer }
:ReturnType:        Manual
:End:


:Begin:
:Function:          resetWaveform
:Pattern:           ResetWaveform[slotNum_Integer]
:Arguments:         { slotNum }
:ArgumentTypes:     { Integer }
:ReturnType:        Manual
:End:
```

The template file should be quite recognizable to you. This example implements separate functions for setup, start, stop, and reset. Since the digital-to-analog hardware can perform its task without additional intervention once it has been set up and started, having separate functions allows the user to perform other tasks while the waveform is being repeatedly generated.

One item of note: **SetupWaveform** takes an integer list as its waveform parameter. The example assumes that the data will be appropriately massaged into raw binary integers.

9.3.4.2 The C code: GenerateWaveform.c

```
/* This is an example using synchronous waveform
generation routines. A 25 Hz sine wave and a 1.25 kHz
square wave are produced at channels 0 and one
respectively using a LAB-NB/LAB-LC board */

#include <stdio.h>
#include <math.h>
#include "mathlink.h"
#include "MitlUtil.h"
#include "NI_DAQ_MAC.h"


/* hard-coded stuff */
```

```
#define TIMEBS3 /* timebase unit of 100 micro second ->
                   output updated every 5 micro secs */
#define CONT1   /* enable continuous waveform
                   generation */
#define NUM_CHANS 1     /* writing to 1 anolog output
                           channels */


int16 *buffer = NULL; /* this array will contain the
                         waveform */
int16 chan_vector[NUM_CHANS];   /* contains the number
                                   of the anolog output
                                   channel */


void setupWaveform (int16 board, int16 channel,
                    int32 sampleInterval,
                    int sampleBuffer[],
                    int32 sampleCount);
void startWaveform(int16 board);
void stopWaveform(int16 board);
void resetWaveform(int16 board);


void setupWaveform (int16 board, int16 channel,
                    int32 sampleInterval,
                    int sampleBuffer[],
                    int32 sampleCount)
{
int16err;
int32 index;

/* fill the buffer with a copy of the List sent from
Mathematica */
MITL_TRY

if (buffer != NULL)/* free the old buffer if it
                      exists */
{
  free_svector(buffer);
  buffer = NULL;
}

buffer = svector(sampleCount);/* make a new buffer and
                                 fill it */
for (index = 0L; index < sampleCount; index++)
              buffer[index] = sampleBuffer[index];
```

```
      chan_vector[0]= channel;   /* initialize channel
                                    vectors for output
                                    channels */

      /* reset resources */
      WF_Grp_Reset(board);

      /* configure the resources */
      err = WF_Grp_Setup (board, NUM_CHANS, chan_vector,
                          sampleInterval, TIMEBS);
      if (err)
      {
        chkerr("WF_Grp_Setup ",err);
        throw_on_error("Failed in WF_Grp_Setup.");
      }

      /* load the buffer for the first channel */
      err = WF_Load (board, chan_vector[0], buffer,
                     sampleCount, CONT);
      if (err)
      {
        chkerr("WF_Load",err);
        WF_Grp_Reset (board);
        throw_on_error("Failed in WF_Load.");
      }

      MLPutSymbol(stdlink, "Null");

      MITL_RECOVER

      if (buffer != NULL)/* free the old buffer if it
                            exists */
      {
        free_svector(buffer);
        buffer = NULL;
      }

      MLPutSymbol(stdlink, "$Failed");
      MITL_ENDTRY

      return;

      }
```

```
void startWaveform(int16 board)
{
  int16err;

  err = WF_Grp_Start(board);
  if (err)
    chkerr("WF_Grp_Start",err);

  MLPutSymbol(stdlink, "Null");
  return;

}


void stopWaveform(int16 board)
{
  int16err;

  err = WF_Grp_Stop(board);
  if (err)
    chkerr("WF_Grp_Stop",err);

  MLPutSymbol(stdlink, "Null");
  return;

}
void resetWaveform(int16 board)
{
  int16err;

  err = WF_Grp_Reset (board);
  if (err)
    chkerr("WF_Grp_Reset",err);

  MLPutSymbol(stdlink, "Null");
  return;
}
```

The *MathLink* C code is quite simple. Again, a more robust implementation of **SetupWaveform** would return a **Waveform** object that encapsulates its parameters; this is left as an exercise for the reader. The only item of note is the handling of the `IntegerList` sent from *Mathematica*. Recall that such lists mentioned in *MathLink* template code will map onto two arguments in the corresponding C code. Note also that we should neither manipulate the contents of the list in place nor assume that it will persist after completion of the C function that uses it. Therefore, the best approach is to copy it into a buffer that the C code can manage itself.

9.3.4.3 Using `GenerateWaveform`

Let us create 1000 points sampled from a Bessel function, a quasi-periodic function of reals.

In:

```
waveform = N[Table[BesselJ[3,z], {z,0,10-0.01,0.01}]];
ListPlot[waveform]
```

Out:

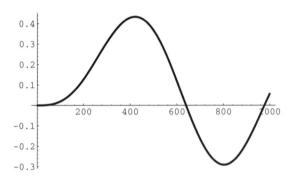

We need to transform the evaluated function's values to appropriate integers. The hardware installed in the author's machine has a 12-bit digital-to-analog converter configured for unipolar output. That is, it expects positive integers from 0 to 4095. The hardware also could be configured for bipolar output (-2048 to 2047), in which case the following transformation would be slightly different.

First, adding **Abs[Min[waveform]]** to each value shifts up every element in **waveform** sufficiently far to ensure that all elements are positive. Then, multiplying by 4095 and applying **Round** transforms the data to integers. This is not a general formula, of course. It is often easiest to use *Mathematica* to quickly visualize data and transform it as needed. Note that, unlike C or FORTRAN, we do not need to have complicated looping constructs: *Mathematica* can operate easily on lists.

In:

```
waveform = Round[4095 * (waveform + Abs[Min
                                       [waveform]])];
```

We now verify that we sensibly transformed the data. If not, we can modify the transformation and reevaluate it.

In:

ListPlot[waveform]

Out:

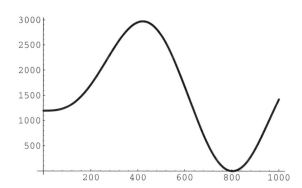

The transformed waveform looks good. The following **SetupWaveform** statement sends the waveform to the digital-to-analog hardware and informs the board to output it at 1 Hz: "Send waveform to the board in slot 6 and have it output on channel 0. The time interval between samples should be 10 ticks of the clock." Since the clock was hardcoded in the C code at 100 µs, this means that the waveform will take:

```
10 ClockTick/Sample * 100 µs/ClockTick * 1000 Sample
```

We do not need to use *Mathematica* to recognize that this is 1 second! That is, we have a waveform with a 1-Hz repetition rate.

In:

SetupWaveform[6,0,10,waveform]

The next three calls demonstrate that we can start and stop the generation of the waveform as often as we like. We can also reset the hardware if we want to clean up.

In:

```
StartWaveform[6]
StopWaveform[6]
ResetWaveform[6]
```

9.3.4.4 Moving on

You can modify the above example to allow more (or fewer) parameters. You might also set the hardware to output simultaneously different waveforms on different channels.

One good reason to output a synthetic waveform is to stress-test a digital circuit (for example, a digital filter). By generating a synthetic analog waveform, routing it through the circuit of interest, and performing data acquisition on the output from the circuit, you can test out a circuit design. If you design circuits, consider making *Mathematica* a prototype tester for your designs.

9.3.5 Additional projects

We have only scratched the surface of a modern data acquisition library such as NI-DAQ. There are many other features common to this and other libraries. You may want to strike out on your own. Find a problem you care about, figure out what services are provided by your laboratory hardware, and wrap the functions needed in *Mathematica* and C code. The following are some suggestions for your amusement.

9.3.5.1 Counting LEDs

Use the digital input/output (DIO) functions of your hardware to count on a set of 8 or 16 LEDs. Each number generated in *Mathematica* can be sent to (or read from) a so-called "port." The bits of the number correspond to the on/off state of the port's lines.

9.3.5.2 Times Square (block character generator)

This is a DIO project. First, work out on paper or in *Mathematica* a set of 8-by-8 or 16-by-16 grids representing a character set. Design a *Mathematica* function that takes an arbitrary string (your message), and converts it to a list of 8- or 16-bit numbers, with each list corresponding to one column of an 8- or 16-high LED strip. Pump the converted message to the appropriate number of DIO ports and see your message appear in lights. If you are especially ambitious – and have enough LEDs and output ports – you could mimic a modern electronic sign, typical of those at airports and sports arenas. Think about sending a *Mathematica* **DensityPlot** to an array of latched LEDs.

9.3.5.3 Home heating controls

Connect a thermometer to an analog input line, and a safe heater to a DIO line (using the appropriate isolating circuitry). Write a *Mathematica* algorithm that periodically reads the thermometer and decides when to turn on and off the heater to achieve a particular time-dependent temperature profile.

9.3.5.4 Elevator controls

Over 100 years after its invention, elevators ("lifts") are still finicky. Use a photocell, stepper motor, pulleys, box, and tube to build a miniature lift. Use *Mathematica* to create and test algorithms for stopping and starting the lift smoothly. Figure out how to get the lift to line up with the doors without having to creep to a halt or oscillate.

9.3.5.5 Dream away

If you are a working engineer, an experimental scientist, or a student, you probably have a host of projects that can be aided and abetted by using *Mathematica* in your laboratory. Both Wolfram Research and the authors of this book would be greatly pleased if any readers who come up with an elegant application and solution to a laboratory problem would consider submitting it to *MathSource*.

9.4   Utility routines

9.4.1   The template file: `MitlMain.tm`

When building a NI-DAQ example, you need to process the specific template along with `MitlMain.tm` one to get a single `.tm.c` file. This template contains the definition for the `Mitl::mitlerr` message that the *MathLink* C code will send to *Mathematica* in case of an error.

```
/* To launch this program from within Mathematica use:
 *    In[1]:= link = Install["theProgramName"]
 *
 * Or, launch this program from a shell and establish a
 * peer-to-peer connection.  When given the prompt
   Listen on:
 * type a port name.
   ( On Unix platforms, a port name is a number less
      than 65536.  On a Mac it's an arbitrary word.)
 * Then, from within Mathematica use:
 *    In[1]:= link = Install["portname",
                             LinkMode->Connect]
 */

:Evaluate:MitL::mitlerr = "MitL run-time error; `1`."
:Evaluate:numberQ[x_]  := NumberQ[N[x]]
:Evaluate:numMatrixQ[x_]  := MatrixQ[x, numberQ]
```

```
:Evaluate:numVectorQ[x_] := VectorQ[x, numberQ]
:Evaluate:numSquareMatrixQ[x_] := Apply[SameQ,
                        Dimensions[x]] && numMatrixQ[x]

#define MITLANSI
#include "MITLUTIL.H"
#include "mathlink.h"

#if !WINDOWS_MATHLINK

int main(int argc, char *argv[])
{
  return MLMain(argc, argv);
}

#else

int PASCAL WinMain( HANDLE hinstCurrent, HANDLE
    hinstPrevious, LPSTR lpszCmdLine, int nCmdShow)
{
  char  buff[512];
  char FAR * argv[32];
  int argc;

  if( !MLInitializeIcon( hinstCurrent, nCmdShow))
    return 1;
  argc = MLStringToArgv( lpszCmdLine, buff, argv, 32);
  return MLMain( argc, argv);
}

#endif

MitlUtil.h
#ifndef _MITL_UTILS_H_
#define _MITL_UTILS_H_

void chkerr(char *s, short err);

#include <setjmp.h>
extern jmp_bufMITLjmp_buf;
extern intsetjmpCalled;

/* convenience macros for error recovery - use as
follows */
/*MITL_TRY */
```

```
/*MITLsomerecipe();*/
/*MLPutxxxxx();*/
/*MLPutxxxxx();*/
/*...*/
/*MITL_RECOVER*/
/*MLPutSymbol(stdlink, "$Failed");*/
/*MITL_ENDTRY*/

#define MITL_TRYif (!setjmp(MITLjmp_buf)){\
setjmpCalled = 1;{
#define MITL_RECOVER}} else{{\
MLClearError(stdlink);
#define MITL_ENDTRY}setjmpCalled = 0;}

void mitlerror(char error_text[]);
void throw_on_error(char error_text[]);
float *vector(long n);
short *svector(long n);
int *ivector(long n);
unsigned char *cvector(long n);
unsigned long *lvector(long n);
double *dvector(long n);
void free_vector(float *v);
void free_svector(short *v);
void free_ivector(int *v);
void free_cvector(unsigned char *v);
void free_lvector(unsigned long *v);
void free_dvector(double *v);

#endif /* _MITL_UTILS_H_ */
```

9.4.2 The C code: MitlUtil.c

This file contains the memory allocation and error routines for the programs
in this chapter. Allocated memory is added to a list so that it can be freed
automatically when there is an error, greatly simplifying the logic of the
examples.

```
#include <stdio.h>
#include <stddef.h>
#include <stdlib.h>
#include <setjmp.h>
#include "mathlink.h"
```

```c
#include "MitlUtil.h"
#define FREE_ARG char*

/* used to send messages via MathLink */
#define ERRBUFSIZE  200
#define MAXMESSAGEPARTS  4
static char *msgv[MAXMESSAGEPARTS];

/* used so we can clean up allocated memory after
throw_on_error */
typedef struct s_plist { void *p; struct s_plist
                              *next; }
plist;
static plist *MITL_memorylist=NULL;

/* used for error recovery */
jmp_buf  MITLjmp_buf;
int  setjmpCalled = 0;

/******************* pointer lists
********************************/

static void insert_plist(plist **head, void *obj)
{
  plist *newElement;

  newElement = (plist *) malloc(sizeof(plist));
  if (!newElement) {
    free(obj);
    mitlerror("allocation failure in insert_plist()");
  }
  newElement->p = obj;
  newElement->next = *head;
  *head = newElement;
}

static int query_plist(plist *head, void *obj)
{
  plist *pl;

  for (pl = head; pl != NULL; pl = pl->next)
    if (pl->p == obj)
      return 1;
  return 0;
}
```

```
static void remove_plist(plist **head, void *obj)
{
  plist *which = *head, *tmp;

  if (obj == which->p) {
    *head = which->next;
    free(which);
  } else {
    while (which->next != NULL && which->next->p != obj)
      which = which->next;
    if (which->next->p == obj) {
      tmp = which->next;
      which->next = which->next->next;
      free(tmp);
    }
  }
}


static void free_plist(plist **head)
{
  plist *tmp;
  while (*head != NULL) {
    tmp = *head;
    *head = tmp->next;
    free(tmp);
  }
}


/************** error handling ****************/
static int send_message_to_mathematica(char *symb, char
*tag, int msgc, char *msgv[])
{
  int pkt,i;

  MLPutFunction( stdlink, "EvaluatePacket", 1);
    MLPutFunction( stdlink, "Message", msgc+1);
      MLPutFunction( stdlink, "MessageName", 2);
        MLPutSymbol( stdlink, symb);
        MLPutString( stdlink, tag);
      for (i = 0; i < msgc; i++) MLPutString( stdlink,
msgv[i]);
  MLEndPacket( stdlink);

  while( (pkt = MLNextPacket( stdlink)) && pkt !=
RETURNPKT)
```

```
      MLNewPacket( stdlink);
   MLNewPacket( stdlink);

   return MLError( stdlink) == MLEOK;
}

void mitlerror(char error_text[])
/* Mathematica in the Laboratory standard error handler
*/
/* This handler uses MathLink to */
/* return its error message. It also returns $Failed */
/* and closes the link. */
{
  char errBuf[ERRBUFSIZE];
  sprintf(errBuf, "%s",error_text);
  msgv[0] = errBuf;
  send_message_to_mathematica("MitL", "mitlerr", 1,
msgv);
}

void throw_on_error(char error_text[])
{
  mitlerror(error_text);
  free_plist(&MITL_memorylist);
  if (setjmpCalled)  {
    setjmpCalled = 0;
    longjmp(MITLjmp_buf, 1);
  }
  else  {
    MLClearError(stdlink);
    MLPutSymbol(stdlink, "$Failed");
    MLEndPacket(stdlink);
    MLNewPacket(stdlink);
    MLClose(stdlink);
    exit(1);
  }
}

/******************** memory allocation
************************/

float *vector(long n)
/* allocate a float vector with n elements */
{
  float *v;
```

```
  v=(float *)malloc((size_t) ((n)*sizeof(float)));
  if (!v) throw_on_error("allocation failure in
vector()");
  insert_plist(&MITL_memorylist, v);
  return v;
}

short *svector(long n)
/* allocate a short vector with n elements */
{
  short *v;

  v=(short *)malloc((size_t) ((n)*sizeof(short)));
  if (!v) throw_on_error("allocation failure in
svector()");
  insert_plist(&MITL_memorylist, v);
  return v;
}

int *ivector(long n)
/* allocate an int vector with n elements */
{
  int *v;

  v=(int *)malloc((size_t) ((n)*sizeof(int)));
  if (!v) throw_on_error("allocation failure in
ivector()");
  insert_plist(&MITL_memorylist, v);
  return v;
}

unsigned char *cvector(long n)
/* allocate an unsigned char vector with n elements */
{
  unsigned char *v;

  v=(unsigned char *)malloc((size_t) ((n)*sizeof
(unsigned char)));
  if (!v) throw_on_error("allocation failure in
cvector()");
  insert_plist(&MITL_memorylist, v);
  return v;
}

unsigned long *lvector(long n)
```

```
                 /* allocate an unsigned long vector with n elements */
                 {
                   unsigned long *v;

                   v=(unsigned long *)malloc((size_t) ((n)*sizeof(long)));
                   if (!v) throw_on_error("allocation failure in
                 lvector()");
                   insert_plist(&MITL_memorylist, v);
                   return v;
                 }

                 double *dvector(long n)
                 /* allocate a double vector with n elements */
                 {
                   double *v;

                   v=(double *)malloc((size_t) ((n)*sizeof(double)));
                   if (!v) throw_on_error("allocation failure in
                 dvector()");
                   insert_plist(&MITL_memorylist, v);
                   return v;
                 }

                 void free_vector(float *v)
                 /* free a float vector allocated with vector() */
                 {
                   remove_plist(&MITL_memorylist, (FREE_ARG) (v));
                   free((FREE_ARG) (v));
                 }

                 void free_svector(short *v)
                 /* free a float vector allocated with vector() */
                 {
                   remove_plist(&MITL_memorylist, (FREE_ARG) (v));
                   free((FREE_ARG) (v));
                 }

                 void free_ivector(int *v)
                 /* free an int vector allocated with ivector() */
                 {
                   remove_plist(&MITL_memorylist, (FREE_ARG) (v));
                   free((FREE_ARG) (v));
                 }

                 void free_cvector(unsigned char *v)
```

```
/* free an unsigned char vector allocated with cvector()
*/
{
  remove_plist(&MITL_memorylist, (FREE_ARG) (v));
  free((FREE_ARG) (v));
}

void free_lvector(unsigned long *v)
/* free an unsigned long vector allocated with lvector()
*/
{
  remove_plist(&MITL_memorylist, (FREE_ARG) (v));
  free((FREE_ARG) (v));
}

void free_dvector(double *v)
/* free a double vector allocated with dvector() */
{
  remove_plist(&MITL_memorylist, (FREE_ARG) (v));
  free((FREE_ARG) (v));
}
```

9.5   References

National Instruments Corporation, "NI-DAQ® Software Reference Manual for Macintosh (Version 4.7)," National Instruments Corporation, Austin, Texas, USA, 1991 and 1994.

Press, W. H., Flannery, B. P., Teukolsky, S. A., Vetterling, W. T., "Numerical Recipes in C (second edition)," Cambridge University Press, Cambridge, United Kingdom, 1992.

CHAPTER 10

Interface hardware design

Some sensors (or other devices) from which your laboratory equipment acquires data will output electrical signals that will be compatible with the data acquisition card in your computer. For example, you can easily obtain solid-state temperature sensors that are powered from a 5-V DC supply and that output a voltage proportional to the absolute temperature in Kelvin, with a scale of 10 mV/K. Thus a temperature of 0 °C would result in a signal of 2.73 V DC. If your computer's analog input card has a 12-bit converter that accepts input signals from 0.000 V DC to +4.096 V DC, then your sensor and card are quite compatible. You should be able to measure temperatures from −273 °C to +135 °C with a resolution of 0.1 K. In practice, the measurable range will be smaller (being limited by the sensor's capability, typically 0 °C to 100 °C), and the accuracy of the sensor will limit useful resolution to about 1 K – but the system performance will still be limited by the sensor rather than the interface card.

If the sensor's output signal is not compatible with the interface card, then you need to design and build an electronic buffer to modify the signal so that it becomes compatible. The electronic buffer may be quite simple – perhaps a completely passive attenuator – or it might be more complicated.

For example, if a temperature sensor had an output 100 mV/K, and we wanted to measure temperatures up to 100 °C, then we would need to divide (attenuate) the sensor's output by a factor of just over 7.6. Such attenuation would reduce the resolution to about 13 mV/K. On the other hand, if the sensor's signal ranged from 5.00 V to 6.00 V as the temperature rose from 0 °C to 100 °C, we would want to reduce the signal by 5.00 V and multiply the resulting difference by, say, 4, thus giving a range of 0.00 V to 4.00 V which is better-matched to our interface card.

You can use *Mathematica* to help you design interface circuitry. In this chapter we look at a few simple electronic buffers and use *Mathematica* to solve the design equations where required. However, the art of matching sensors to measuring devices is quite a large subject in itself, and here it is possible for us only to cover a few examples in fleeting detail. Horowitz &

Hill (1980) and Loxton & Pope (1986) both give good introductions to the subject from the hardware perspective; Riddle & Dick (1994) give a more in-depth treatment of circuit design (including filters and noise parameters) with *Mathematica*. We end this chapter with a brief look at noise, and we describe some methods for noise reduction.

10.1 Simple buffer circuits

10.1.1 Attenuators

The simplest signal modifier is a T attenuator (Fig. 10a). In the symmetric T attenuator, the input and output impedances are identical in value, **z**, and the signal is attenuated by a factor **a**. By applying Kirchhoff's laws to the circuit, we obtain a set of equations that can be solved by *Mathematica*.

In:

```
Solve[{i1 z == i1 r1 + i2 r2,
       i2 r2 == i3 r1 + i3 z,
       i1 z == i1 r1 + i3 r1 + i3 z,
       i1 == i2 + i3,
       i3 == a i1},
      {r1,r2},{i1,i2}]
```

Out:

$$\left\{\left\{r2 \to \frac{-2\,a\,z}{(-1 + a)\,(1 + a)},\; r1 \to \frac{(1 - a)\,z}{1 + a}\right\}\right\}$$

By replacing **z** with our chosen impedance, 600 Ω, and **a** with an attenuation factor of one-half, we can calculate values for **r1** and **r2**.

In:

```
% /. {z->600, a->0.5}
```

Out:

```
{{r2 -> 800., r1 -> 200.}}
```

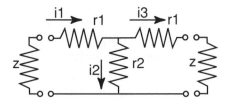

Figure 10a T attenuator

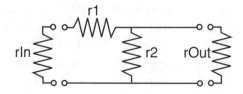

Figure 10b L attenuator

Another attenuator design is known as the L configuration (Fig. 10b). Where you want to match, say, a low-impedance 50 Ω coaxial cable to a higher-impedance amplifier, and where some attenuation can be tolerated, you can use an L attenuator. The design process involves finding resistor values such that both loads see a circuit impedance equal to their own. Although we have not printed out the result of **Solve**, by so doing you can discover conditions that would result in the resistors having (unrealizable) complex values, and so map out forbidden regions of circuit performance.

In:

```
Solve[{rIn==r1+1/(1/r2+1/rOut),
        rOut==1/(1/r2+1/(r1+rIn))},{r1,r2}];
% /. {rIn->200, rOut->50}//N
```

Out:

```
{{r1 -> -173.205, r2 -> -57.735},
  {r1 -> 173.205, r2 -> 57.735}}
```

(Note that the solution involves taking square roots, and so the values for the resistors can be positive or negative, as far as the mathematics is concerned.)

10.1.2 Amplifiers

10.1.2.1 Inverting-summing amplifier

If your signal is not matched in magnitude to the input range of the computer's interface card, then Fig. 10c shows an operational amplifier circuit that will cure the problem.

Note that this circuit has two inputs. By applying a stable DC voltage at one input and your signal at the other, you will be able to remove any DC plateau present in your signal if the amplifier's gain is unity. (Remember to verify the expected output voltage for the full range of your input signal.) Note that the impedance of the DC voltage applied to the nonsignal input should be ten or more times smaller than its input resistor (R_1 or R_2).

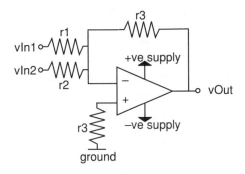

Figure 10c Two-input amplifier

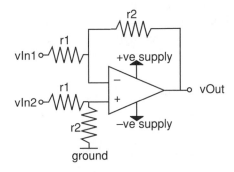

Figure 10d Differential amplifier

The output voltage of the circuit is given by

$$v_{\text{Out}} = -R_3 \left(\frac{v_{\text{In1}}}{R_1} + \frac{v_{\text{In2}}}{R_2} \right)$$

If your signal is AC, you can insert a DC-blocking capacitor between the signal source and its input resistor (R_1 or R_2). The reactance of the capacitor should be at least ten times smaller than the input resistor at the lowest frequency you intend to use. The function **reactanceC** calculates the reactance in ohms, of a capacitor of **c** farads at frequency **f** hertz.

In:

```
reactanceC[c_,f_]:=1/(2 Pi f c)//N;
reactanceC[10^-6, 1000]
```

Out:

```
159.155
```

10.1.2.2 Differential amplifier

If you are trying to measure a small difference between two signals, then you may need to use a differential amplifier (Fig. 10d).

Table 10:i Solution circuits for signal-handling problems

Problem	Solution circuit
Unipolar signal has wide dynamic range	Logarithmic amplifier
Signal has sharp peaks to measure or is only present occasionally	Peak-hold circuit
Unipolar version of bipolar signal required	Op-amp rectifier
Phase comparison of two AC signals required	Phase-sensitive detector
Signal change-rate required	Differentiator
Signal sum-to-date required	Integrator
Fixed-frequency signal in noise	Phase-locked loop

The output voltage, v_{Out}, is:

$$v_{Out} = \frac{R_2}{R_1}(v_{In1} - v_{In2})$$

For this circuit, the two paired resistors (that is, both resistors labeled R_1) must be well matched – possibly to the 0.1% level or better. You can buy such high-precision resistors from specialist electronic suppliers.

10.1.2.3 Going on

There are many useful signal adjusting circuits that are built around operational amplifier chips. Table 10:i indicates circuits that you might use to help tackle a range of signal-related problems. For information on their implementation see, for example, Horowitz & Hill (1980) or Millman & Halkias (1972).

10.2 Noise control

In this section we discuss something that nobody wants: noise. We often think that there is a connection between noise in data and poppies in an ornamental garden's rose bed. Why? The term "noise" is used commonly to refer to any signal in which we are uninterested, regardless of whether the signal is the result of a random physical process (true noise?) or the breakthrough of a signal from another data channel (someone else's signal?). So one experimenter's noise might be another's data, just as our poppies are perfectly good plants that are not supposed to be in a rose bed. Here we discuss noise and design a few circuits that you can use to help limit the effect of noise in your data.

10.2.1 Noise in general

Qualitatively, noise needs no introduction. The hiss you hear in a radio receiver or the snowy picture on a television when off-channel are familiar to us all. True noise, as distinct from the breakthrough of other signals, is categorized into three main types.

Thermal noise, or Johnson noise, is caused by the Brownian motion of electrons within a conductor. Because the random movement of some of the electrons toward one edge of a conductor is not balanced exactly by a similar number of electrons moving to the opposite edge, a net voltage is generated momentarily between the edges. This voltage is present even when the conductor is not connected into a circuit. Of course, the mean value of the noise voltage is zero. For a conductor with resistance R, the thermal-noise-generated voltage squared is given by the equation $V^2 = 4kTR\Delta f$ where k is Boltzmann's constant (1.38×10^{-23} JK^{-1}), T is the temperature in Kelvin, and Δf is the bandwidth (in hertz) used for the measurement.

The bandwidth is important because thermal noise is white; that is, it has equal power per hertz, regardless of whether we measure its power with the bandwidth centered at 10 Hz or at 20 MHz. So the mean squared voltage measured with a bandpass from 40 kHz to 80 kHz will be twice that measured from 10 kHz to 90 kHz, and the same as that measured from 10.030 MHz to 10.070 MHz.

As an indication of amplitude, a 10-kΩ resistor at room temperature has a root-mean-squared voltage of 2.6 μV when measured with a bandpass from DC to 40 kHz; a 1-MΩ resistor measured with a band-pass from DC to 1 MHz would have a root-mean-squared voltage of 0.13 mV.

In:
```
thermalVrms[r_,t_,df_]:=Module[{k=1.38 10^-23},
                        Sqrt[4 k r t df]];
thermalVrms[10^6,300,10^6]
```
Out:
```
0.000128686
```

The second main source of white noise is called "shot noise" and is caused by the Poisson distribution of electrons flowing through any conductor. Although, on average, exactly the same number of electrons pass along a section of conductor in any period of time, the actual number of electrons varies with a Poisson distribution. In some samples there will be more electrons than in others, so the current passed appears to vary. The mean-square current variation is given by the equation $I_{\mathrm{ms}}^2 = 2eI\Delta f$, where e is the electronic charge $(1.6 \times 10^{-19}\,\mathrm{C})$, I is the mean current in amperes, and Δf is the bandwidth (in hertz) used for the measurement. A mean current of $1\,\mu\mathrm{A}$, measured with a bandpass from DC to $1\,\mathrm{MHz}$ has a root-mean-square variation of $0.6\,\mathrm{nA}$.

These two white-noise sources are well understood from a theoretical point of view; Carlson (1981), Connor (1982), Delaney (1969), and Horowitz & Hill (1983) make good reading if you want to learn more about noise and how it can limit your experimental accuracy.

The third main source of noise is called "flicker noise" and its source is less well understood (see Horowitz & Hill (1983) and Takayasu (1990)). Unlike thermal and shot noise, flicker noise is not white – and is often referred to as pink – because it has higher power at lower frequencies, just as a pink object reflects proportionally more longer-wavelength (red) light than a white object.

Flicker noise can be caused by, for example, faults in the crystal lattice within semiconductor devices or by discontinuities in deposited conductive films on resistors. Its spectrum contains an equal power per octave or decade of frequency. So the root-mean-square noise voltage in a bandwidth DC to $10\,\mathrm{Hz}$ will be the same as that from $10\,\mathrm{Hz}$ to $100\,\mathrm{Hz}$, despite the bandwidth being ten times larger. For a white-noise source, on the other hand, the noise power would have multiplied ten times. Hence, flicker noise is of principal importance at lower frequencies; however, because it is variable even between two transistors made in the same production run, predicting its magnitude is very difficult, if not impossible.

Let us now look at ways of reducing the effects of noise.

10.2.2 Reducing true noise

Both thermal noise and shot noise can be reduced by keeping the bandwidth of your data acquisition system as small as possible, while still satisfying Nyquist's sampling theorem, of course. For example, if you are trying to capture information at high frequency, f, you might not need a bandwidth from DC to f. A lower-noise option might be to filter the incoming signal to greatly attenuate the noise outside your spectral region of interest. You might also consider heterodyning the incoming signal with a local oscillator and then carrying out acquisition and analysis in a lower-frequency band.

The subject of filter design is large and outside our scope for this book. Millman & Halkias (1972), Connor (1986), Horowitz & Hill (1980), Northrop (1990), and Riddle & Dick (1994) offer further reading.

Thermal noise can be reduced by lowering the temperature of the noise-creating components or by reducing the resistance of the sensor, if possible.

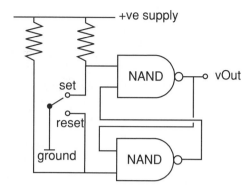

Figure 10e Switch debouncer

10.2.3 Switch debounce

One of the simplest forms of noise is called "switch bounce." It is not related to the forms of noise we have discussed above, but it is a major nuisance in digital systems. Switch bounce is caused when the mechanically fine contacts within the switch strike each other and then recoil a number of times, just as a dropped tennis ball bounces several times before it finally settles on the ground. If you are using a mechanical switch to input data to an interface card, then you do not want each bounce of the contacts registering: you probably just want a single level-change on each activation of the switch.

The circuit in Fig. 10e achieves this by latching on the first bounce of any switch activation (in either direction). Try working through the circuit with logic 0 and 1 states, remembering that the bounce phenomenon affects only the "make" or destination side of the switch.

The value of resistors will depend on the type of gates that you are using. Typical values might be 1 kΩ for TTL-LS gates and 10 kΩ for CMOS gates.

10.2.4 Level detection

Another case of noise control is where a nondithering digital output is required from an analog signal. When the analog signal rises above some voltage, the circuit switches to one state; when the analog signal falls below some voltage, the circuit switches to the opposite state. A typical example might be a thermostat: When the sensor's temperature falls, its output voltage drops, and the heater is switched on.

In theory, you could implement this with a voltage comparator as shown below. However, any noise present on the output of the temperature sensor will cause the output of the switching circuit to jitter between on and off, possibly causing a lot of interference to nearby circuitry. An operational-amplifier may have a gain of 10^4, so the change in input voltage that will cause a swing from, say, -5 to $+5$ V is only 1 mV!

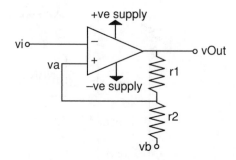

Figure 10f Schmitt trigger

The circuit shown in Fig. 10f solves this problem by adjusting the threshold at which the circuit changes state. This circuit is known as a Schmitt trigger. With V_i less than V_a, the output is high, and so the current flowing from the output through R_1 keeps the threshold voltage higher than if R_1 were not present. When V_i rises above V_a, the output changes state, and the feedback through R_1 now lowers the threshold at which the circuit will return to a low state. The difference in threshold voltages prevents the circuit from jittering in response to small amounts of noise on the input.

The equations that describe the high-low and low-high transition voltages are respectively

$$V_{HL} = \frac{R_1 V_b}{R_1 + R_2} + \frac{R_2 V_o}{R_1 + R_2}$$

and

$$V_{LH} = \frac{R_1 V_b}{R_1 + R_2} - \frac{R_2 V_o}{R_1 + R_2}.$$

We can use *Mathematica* to solve these equations, providing us with values for the various components, if we can specify V_{HL}, V_{LH}, and some of the component values and circuit voltages. You can tackle the design of the circuit in a number of ways. The two functions **vhl** and **vlh** return V_{HL} and V_{LH} given resistor and voltage values; try plotting the thresholds to get a feel for the circuit operation.

In:

```
vhl[r1_,r2_,vb_,vo_]:=
    Module[{},((r1 vb)/(r1+r2))+((r2 vo)/(r1+r2))//N];
vlh[r1_,r2_,vb_,vo_]:=
    Module[{},((r1 vb)/(r1+r2))-((r2 vo)/(r1+r2))//N];

vb=2.0;
Plot[{vhl[10000,rB,vb,5],
    vlh[10000,rB,vb,5]},{rB,10,1000},
    AxesLabel->{"rB","thresholds V"}];
```

Out:

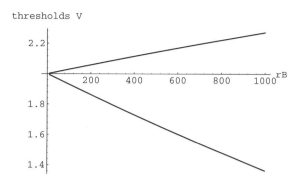

Another Schmitt-trigger design is solved with *Mathematica* by Riddle & Dick (1994).

10.3 Signal generator

When you are testing data acquisition equipment, it is useful to have a signal source with which to test both the equipment and your software. Fig. 10g shows a simple circuit that generates five different voltage outputs.

Point A is fed by a constant-voltage generator chip (a Texas Instruments TLE2425) and is at nominally 2.50 ± 0.03 V; the current drawn through Point A should not exceed a few milliamps. You should use a calibrated bandgap voltage reference if you require higher accuracy.

Point B samples the output of an LM335Z temperature sensor (SGS-Thomson). You should not draw more than a few tens of microamps from this point. The voltage at Point B is the device's temperature in Kelvin divided by 100, so an ambient temperature of 20°C produces a reading of 2.93 V.

Point C samples the emitter of a MEL-12 Darlington phototransistor; as the ambient light level increases, so will the voltage at Point C.

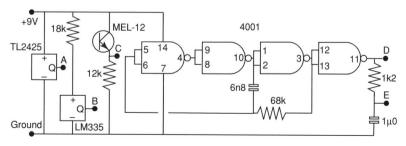

Figure 10g Simple signal generator

A CMOS 4001 quad NAND gate chip is used to generate a ~1200-Hz square wave with ~60% duty-cycle at Point D. Point E gives a filtered version of the square wave – a sawtooth wave. You can alter the frequency at these two points by changing the values of the 6.8-nF capacitor or the 68-kΩ resistor. The 1.2-kΩ resistor and the 1-μF capacitor control the extent of filtering and the voltage range of the sawtooth wave.

10.4 References

Carlson, A. B., "Communication Systems," McGraw-Hill International, London, United Kingdom, 1981.

Connor, F. R., "Noise," Edward Arnold, London, United Kingdom, 1982.

Connor, F. R., "Networks," Edward Arnold, London, United Kingdom, 1986.

Delaney, C. F. G., "Electronics for the Physicist," Penguin Books, Harmondsworth, United Kingdom, 1969.

Horowitz, P., Hill, W., "The Art of Electronics," Cambridge University Press, Cambridge, United Kingdom, 1980.

Loxton, R., Pope, P., "Instrumentation — a Reader," Open University Press, Milton Keynes, United Kingdom, 1986.

Millman, J., Halkias, C. C., "Integrated Electronics: Analog and Digital Circuits and Systems," McGraw-Hill, Tokyo, Japan, 1972.

Northrop, R. B., "Analog Electronic Circuits: Analysis and Applications," Addison-Wesley, Reading, MA, USA, 1990.

Riddle, A., Dick, S., "Applied Electronic Engineering with *Mathematica*," Addison-Wesley, Reading, MA, USA, 1994.

Takayasu, H., "Fractals in the Physical Sciences," Manchester University Press, Manchester, United Kingdom, 1990.

Appendix

A.1 Defining your own functions: a brief summary

You will often find that you need to define your own functions (rather like writing subroutine, function, or procedure code in other computer languages). The simplest way to write your own function is to use **Module**, which allows you to have local variables (that is, variables that are not visible outside the scope of **Module**'s declaration). *Mathematica*'s patternmatching parser provides vetting both of argument types and of the number of arguments. (Function overloading, as available in C++ for example, is permitted.)

A typical implementation of a function is **power**: it raises the first argument to an integer power and returns the product of that result and **a**.

In:

```
power3[x_,y_Integer]=Module[{a=3},
                           a x^y];
power3[2,6]
```
Out:

```
192
```

The symbol **x_** means "a symbol called **x**." Specifying a type after the underscore restricts that variable to be of the specified type: **y** must be an integer in **power3**. Any attempt to call **power3** with a non-integer second argument will result in the function being returned unevaluated.

In:

```
power3[2,6.1]
```
Out:

```
power3[2, 6.1]
```

If we now define another function **power3** without the integer restriction (but which, for our illustrative purposes, has a different value of **a**), *Mathematica* will use the appropriate version.

In:

```
power3[x_,y_]=Module[{a=33}, a x^y];
power3[2,6.1]
```

Out:

```
2263.59
```

In:

```
power3[2,6]
```

Out:

```
192
```

In:

```
power3[2, 6.0]
```

Out:

```
2112.
```

Checking function arguments and overloading functions are useful for ensuring that the code within a function is passed the correct types of data. If you are building a large application, these features can help you to distribute code production among different programmers and to gradually bring code into use as versions of functions become available (incremental development.)

We strongly recommend that you read the sections in the *Mathematica* manual that describe function writing, pattern matching, and variable scope lifetimes. The scope of variables in *Mathematica* resembles that of a block-structured language like Pascal or Algol; the differences between the structures built around **Module** and **Block** require careful attention.

A.2   Pure anonymous functions

Functions in mathematics always have a name, such as "Sine" or "Bessel", because we need a tag by which many people can identify the relevant mathematical operation. In general, computer languages refer to functions by either their common names or contractions thereof. For example, sine becomes `sin` and the square-root function becomes `sqrt` in C and many other computer languages.

If you want to carry out some operation on each member of a dataset, *Mathematica*'s **Map** function will map a function that carries out the operation over all members of a list. **Map** is convenient because it saves you from having to write a looping structure that does the same task. For example, if you have a list of numbers and you want their squareroots, you could write

In:

```
data={1., 2., 3., 4., 5., 6., 7.};
result={};
For[i=1, i<=Length[data], i++,
    AppendTo[result, Sqrt[data[[i]] ] ];
    ];
result
```

Out:

```
{1., 1.41421, 1.73205, 2., 2.23607, 2.44949, 2.64575}
```

or you could use **Map** to take the (function) **Sqrt** of each of the dataset members:

In:

```
result=Map[Function[Sqrt[#]],data]
```

Out:

```
{1., 1.41421, 1.73205, 2., 2.23607, 2.44949, 2.64575}
```

By using **Map**, you do not have to be concerned about the process of repetitively operating on the dataset; **Map** does all that for you.

However, we do have to specify a place where each member of the dataset will become the function's argument – and we do so with the **#** symbol. The need to specify the position of the argument is more apparent if we use a function that takes two arguments. For example, the function **Log[b,z]** returns the base-**b** logarithm of **z**; calculating the base-10 logarithm of the dataset members is quite different from calculating the logarithm of 10 in the base of the members of data.

In:

```
Map[Function[Log[10,#]], data]
```

Out:

```
{0., 0.30103, 0.477121, 0.60206, 0.69897, 0.778151,
0.845098}
```

In:

```
Map[Function[Log[#,10]], data]
```

Out:

```
{ComplexInfinity, 3.32193, 2.0959, 1.66096, 1.43068,
1.2851, 1.18329}
```

Although **Map**'s syntax may seem long-winded, it is possible to use the short infix form of **Map** and the postfix form of **Function**. The postfix form of **Function** is **&**; you place an **&** at the end of the code that you want to declare as a function for **Map** to use. The infix form of **Map** is **/@**; you place an **/@** between the declared function and the data over which the mapping will take place. Confused? Here is the shortened code:

In:

```
result= Sqrt[#]& /@ data
```

Out:

```
{1., 1.41421, 1.73205, 2., 2.23607, 2.44949, 2.64575}
```

Although it may be less readable if you are new to *Mathematica*, it is shorter and, perhaps, more directly related to the sequence of operations. (Most computer languages have notation that is obscure, at first sight; familiarity helps to clear obfuscation.)

We can apply an operation to members of a dataset even if the function is not one of *Mathematica*'s built-in functions. For example, if we declare a function **myFunc** that raises its argument to the 3/2 power, then we can map **myFunc** over data, just as we did with **Sqrt**.

In:

```
myFunc[x_]:=x^(3/2);
result=myFunc[#]& /@ data
```

Out:

```
{1., 2.82843, 5.19615, 8., 11.1803, 14.6969, 18.5203}
```

If we were never to use **myFunc** again, its declaration would be rather wasteful. Then why even give the function a name? After all, **myFunc** is not going to be meaningful to anyone else – and in a few days, probably not even to ourselves! If we were to do away with the function's name, we might write

In:

```
result=#^(3/2)& /@ data
```

Out:

```
{1., 2.82843, 5.19615, 8., 11.1803, 14.6969, 18.5203}
```

Thus we have created an anonymous – or pure – function. The function "something raised to the power 3/2" exists only within the scope of the mapping operation on that one line of code. If we want to use the same functionality elsewhere, we can either lazily cut-and-paste the anonymous function or, more rigorously, declare it properly as a named function.

Anonymous functions can operate on different members of a structured argument. Below, our anonymous function operates on a pair of numbers; we sum the first number and the squareroot of the second number in each pair. The symbol **#[[1]]** means "take the first member of the argument."

In:

```
dataPairs={{1,2},{2,3},{3,4}};
(#[[1]]+Sqrt[#[[2]]])& /@ dataPairs
```

Out:

```
{1 + Sqrt[2], 2 + Sqrt[3], 5}
```

If you have problems with using a simple anonymous function, it is usually because there is a mistake with the form of the argument. Try printing out the first element of the list over which you are trying to map your function.

Anonymous functions can also operate on multiple arguments. In the previous example, we worked on a single argument that was a two-element list. If the arguments are to be treated as individuals, then we need to note a change of syntax. Just as multiple arguments for, say, the **Log** function are enclosed within a sequence delimited by square brackets, such as **Log[10,3]**, so we must use multiple arguments for our anonymous function. We refer to the **n**-th argument using the notation **#n**.

In:

```
(#1+Sqrt[#2])& [aa,bb]
```

Out:

```
aa + Sqrt[bb]
```

By itself, the ability to take multiple arguments is not very useful. We cannot use a two-argument function alone because we cannot declare a list of the form **{[1,2], [2,3], [3,4]}**; it is syntactically incorrect, and *Mathematica*'s command parser will reject it with an accompanying error message.

We can use multiple-argument functions by attaching a symbolic name to them.

In:

```
myNewFunc=(#1+Sqrt[#2])&;
```

We can then write

In:

```
myNewFunc[2., 3.]
```

Out:

```
3.73205
```

You might (correctly) think that this is rather like declaring a named function in the more traditional manner.

For somewhat advanced levels of programming, such a feature does provide flexibility in assignation of functionality to a function name and gives program design advantages akin to function pointers in C.

In:

```
myCode[x_,y_,f_]:=Module[{}, f[x,y]];
myCode[2., 3., myNewFunc]
```

Out:

```
3.73205
```

In:

```
myTooNewFunc=(#1^3+Sqrt[#2])&;
myCode[2., 3., myTooNewFunc]
```

Out:

```
9.73205
```

Index